Infotech

English for computer users

Fourth Edition

Student's Book

CAMBRIDGE
UNIVERSITY PRESS

Santiago Remacha Esteras

CAMBRIDGE UNIVERSITY PRESS
Cambridge, New York, Melbourne, Madrid, Cape Town, Singapore, São Paulo, Delhi

Cambridge University Press
The Edinburgh Building, Cambridge CB2 8RU, UK

www.cambridge.org
Information on this title: www.cambridge.org/9780521702997

First published 2008
Reprinted 2008

Printed in Dubai by Oriental Press

A catalogue record for this publication is available from the British Library

ISBN 978-0-521-702997 Student's Book
ISBN 978-0-521-703000 Teacher's Book
ISBN 978-0-521-703017 Audio CD

Contents

	UNIT	LISTENING	READING
Module 1 **Computers today**	**1** Living in a digital age	Computers at work	The digital age The magic of computers
	2 Computer essentials	Different types of computer	Advertising slogans What is a computer?
	3 Inside the system	A PC system	Technical specifications What is inside a PC system? How memory is measured
	4 Buying a computer	In a computer shop Choosing the right computer	Computer adverts Technical specifications
Module 2 **Input/Output devices**	**5** Type, click and talk!	Describing input devices Mouse actions	Interacting with your computer Speech recognition systems
	6 Capture your favourite image	Scanners	The eyes of your computer Press release: a digital camera
	7 Display screens and ergonomics	Choosing the right display device Ergonomics	How screen displays work
	8 Choosing a printer	Multi-function printers	Which type of printer should I buy? Printer adverts
	9 Devices for the disabled	Assistive technologies for the blind	Computers for the disabled
Module 3 **Storage devices**	**10** Magnetic storage	Buying a portable hard drive	Magnetic storage
	11 Optical storage	CDs and DVDs	Optical discs and drives
	12 Flash memory	Flash drives	Memory in a flash!
Module 4 **Basic software**	**13** The operating system (OS)	Windows Vista	GUI operating systems
	14 Word processing (WP)	The Cut and Paste technique	WP tools
	15 Spreadsheets and databases	The Excel spreadsheet program	An invoice and covering letter Databases

SPEAKING	WRITING	LANGUAGE WORK	VOCABULARY
Discussing what computers do	A short summary of a discussion	Collocations 1	Basic computer terms, computers in education, banks, offices, airports, libraries, entertainment, etc.
Describing a diagram	An email explaining the benefits of laptops and tablet PCs	Classifying	Basic hardware and software terminology
Describing your ideal computer system	Notes about your ideal computer system	Defining relative clauses	*Processor, chip, control unit, arithmetic logic unit*, etc. Units of memory: KB, MB, GB, etc.
Role play – buying a computer	An email recommending a computer	Language functions in a computer shop	Vocabulary tree: revision of vocabulary from Module 1
Describing input devices		Describing functions and features	Input/Output devices, groups of keys, mouse actions
Describing a camera		Superlatives Suffixes	Scanners, cameras
Discussing which display devices you would most like to own	Guidelines for an ergonomic school or office	Instructions and advice	Display screens, ergonomics
Choosing the right printer	An email to a friend comparing two printers	Connectors 1 Comparatives	Types of printer, printer technology
Discussing assistive technology	An email summarizing the different assistive technologies available	Noun phrases	Devices for the disabled
Discussing how to protect your data	An email explaining hard drive precautions	Precautions Word building	Types of magnetic storage, technical details of magnetic storage
Choosing storage devices	A post on a forum discussion about format wars	Connectors 2	Types of optical storage, technical details of optical storage
Describing flash drives	A text message to a friend explaining the difference between MP3 and MP4	Word building	Types of flash drive, technical details of flash memory
Comparing user interfaces	A summary of a text	Countable and uncountable nouns Articles	GUIs, the WIMP environment, desktop features, etc.
Giving instructions for carrying out tasks in Word	Instructions for using *Find and Replace* in Word	Giving and following instructions	Functions and features of word processors
Discussing the software you use at home and at work	A fax of complaint	Plurals	Functions and features of spreadsheets and databases

	UNIT	LISTENING	READING
Module 5 Faces of the Internet	**16** The Internet and email	Internet basics	Internet FAQs Email features
	17 The Web	E-commerce and online banking	A typical web page The collectives of cyberspace
	18 Chat and conferencing	At a cybercafé	Virtual meetings Netiquette
	19 Internet security	Safety online for children	Security and privacy on the Internet The history of hacking
Module 6 Creative software	**20** Graphics and design	The toolbox	Computer graphics
	21 Desktop publishing	Steps in a DTP publication	What is desktop publishing? Steps in a DTP publication
	22 Multimedia	Components and system requirements	Multimedia magic!
	23 Web design	Designing a website	Web page design
Module 7 Programming / Jobs in ICT	**24** Program design and computer languages	Steps in programming	Computer languages
	25 Java™	The history of Java	Java applets The Java language
	26 Jobs in ICT	IT professionals A job interview	Job adverts A letter of application
Module 8 Computers tomorrow	**27** Communication systems	VoIP technology	Channels of communication
	28 Networks	Small networks	Networking FAQs
	29 Video games	Present and future trends in gaming	Game genres
	30 New technologies	RFID tags	Future trends

SPEAKING	WRITING	LANGUAGE WORK	VOCABULARY
Discussing the Internet and what you use it for	A reply to an email about the history of the Internet	Questions	Internet basics, internet and email features
Discussing what you use the Web for	An article about internet phenomena	Collocations 2 The prefixes *e-* and *cyber-*	Web basics, web addresses Online shopping and banking
Discussing online chat Planning your own cybercafé and presenting your plans	An online conversation	Chat abbreviations	Online chat and conferencing
Discussing internet issues	A summary of a discussion	The past simple	Internet security, types of internet crime
Choosing graphics software	Describing graphics	The *-ing* form	Types of graphics, the toolbox
A debate: e-publishing vs. paper publishing	A letter to a newspaper	Order of adjectives	Desktop publishing basics
Discussing applications of multimedia	A blog entry about the use of multimedia	Conditional sentences	Multimedia components and features
Discussing blogs	A home page A blog entry	Modal verbs	Aspects and tools of web design
Describing computer languages	Notes from a training course	Word building The infinitive	Programming, computer languages
Discussing your experience with computers		The *-ed* form	Java applets
Discussing the personal qualities needed for certain jobs	A letter of application for a job A CV	*For, since, ago, until* The present perfect	IT professions, professional skills and qualities
Explaining VoIP technology from a diagram Describing and discussing mobile phones	A summary of a discussion for a blog post	The passive	ICT systems, mobile phones
Presenting a description of a network	A description of a network	Phrasal verbs	Types of network, network architecture, network topology
Discussing your favourite games and game platforms Discussing the pros and cons of gaming	An essay: The pros and cons of gaming	Adverbs	Game platforms, game genres
Discussing and comparing predictions	Captions for short texts Predictions	Future forms	Future trends in technology: nanotechnology, AI, biometrics, etc.

Thanks and acknowledgements

The author would especially like to express his gratitude to Paz, Marina and Violeta. My special thanks to Nick Robinson for his invaluable feedback and for editing the typescript and to Tony Garside for his input and expertise. Thanks are also due to the teachers and students of Pilar Lorengar High School, Zaragoza. Thanks to Cambridge University Press for their vision, support and faith in the project, to Matt Robinson for his generous contribution on video games, and to Angel Benedí and Elena Marco for their help and advice.

We are grateful to all the teachers who provided input and advice during the development of this new edition.

The authors and publishers acknowledge the following sources of copyright material and are grateful for the permissions granted. While every effort has been made, it has not always been possible to identify the sources of all the material used, or to trace all copyright holders. If any omissions are brought to our notice, we will be happy to include the appropriate acknowledgements on reprinting.

p. 31 Kodak Easy Share C663 Press Release 3 January 2006, © 2003-2007 Ketchum Inc.; p. 41 adapted text describing Canon SELPHYCP750 Photo Printer, © Canon 2007; p. 41 adapted text describing Vutek UltraVu II 5330 from www.efi-vutek.com/products; p. 41 Brother UK Ltd for text describing Brother HL Network Colour Laser Printer HL 2700 CN, www.brother.com.au; p. 59 Amazon for description of Olympus WS-320M, www.amazon.com, © 1996-2007, Amazon.com, Inc. and its affiliates; p. 67: text adapted from www.linux.org, content available under terms of GFDL License; p. 72: adapted text 'Spellcheckers' and p. 105 adapted article, 'Desktop publishing' reprinted from *Understanding Computers* by Nathan Sheldroff, J Sterling Hutto and Ken Fromm, 1993; p. 84 article adapted from 'Tour the collectives of cyberspace' from BusinessWeek online, 20 June 2005; p. 146; article adapted from 'Game genres', www.ringsurf.com.

The publishers are grateful to the following for permission to reproduce copyright photographs and material:

Key: l = left, c = centre, t = top, b = bottom

ACE STOCK LIMITED/Alamy for p. 2(tl), Ian Shaw/Alamy for p. 2(tr), Wayne Eastep/Getty Images for p. 2(bl), Juliet Brauner/Alamy for p. 2(br), Elizabeth Whiting & Associates/Alamy for p. 5, Mark M. Lawrence/Corbis for p. 8(l), Geri Lavrov/Alamy for p. 8(r), Courtesy of Dell Inc. for p. 11 (l), Tahesi Takahara/Science Photo Library for p. 11(r), Intel Corporation, 2007 for p. 12(bl), PC World for p. 16(t), David Paul Morris/Getty Images for p. 16(c), Profimedia International s.r.o./Alamy for p. 16(b), Photo courtesy of Sun Microsystems, Inc. for p. 18, Gateway, Inc. for p. 19(t), SONY for p. 19(c), Courtesy of Dell Inc. for p. 19(bl), (br), PlayStation and the PlayStation logo, PSP and PLAYSTATION 3 are trademarks or registered trademarks of Sony Computer Entertainment Inc. for p. 23(t), www.CartoonStock.com for p. 25, Source: Nuance Communications for p. 26, Paul Hardy/Corbis for p. 30, © Kodak for p. 31, Image courtesy of NEC Display Solutions for p. 34(tl), Courtesy of Dell Inc. for p. 34(tr), © Cambridge University Press, image courtesy thephotounit for p. 34(cl), with compliments of Pioneer GB Ltd for p. 34(cr), Canon Compact Photo Printer SELPHY CP750 for p. 41(l), Picture courtesy of EFI-Vutek for p. 41(tr), Image of Brother HL-4040CN provided courtesy of Brother Industries UK for p. 41(br), Sally Lancaster/Format Photographers for p. 42(tr), (cl), De Repentigny, Publiphoto Diffusion/ Science Photo Library for p. 42(cr), AbilityNet for p. 42(bl), ImageState/Alamy for p. 43, Textlink 9100Mobile for p. 44(cl), AbilityNet for p. 44(bl), Najlah Feanny/Corbis for p. 46(t), Picture courtesy of: GW Micro, Inc. for p. 46(b), © Iomega for p. 48(b), Photograph courtesy of Toshiba for p. 49, Vadym Kharkivskiy/Alamy for p. 52, SONY for p. 53, p. 55, Photo courtesy of D-Link Coporation for p. 57(tl), SONY for p. 57(tlc), The EDGE Tech Corp for p. 57(tcr), © SanDisk for p. 57(tr), NINTENDO for p. 57(bl), The EDGE Tech Corp for p. 57(br), © SanDisk for p. 58, © Olympus for p. 59, © SanDisk for p. 60(r), © Creative Technology Limited for p. 60(l), Grain Belt Pictures/Alamy for p. 82, vario images GmbH & Co.KG/Alamy for p. 88, Tony Metaxas/Getty Images for p. 89, AFP/Getty Images for p. 90, www.CartoonStock.com for p. 92, Hekimian Julien/ Corbis Sygma for p. 93, Steve Allen/Alamy for p. 101(br), Microsoft Encarta Premium 2007 box shot reprinted with permission from Microsoft Corporation for p. 110, Patrick Steel/Alamy for p. 111, Helen King/Corbis for p. 113(tr), Marcus Mok/Getty Images for p. 113(br), Helen King/Corbis for p. 117, WoodyStock/Alamy for p. 130, Teletext for p. 135(tl), Christoph Rosenberger/Alamy for p. 135(tcr), TomTom GO 720 images for p. 135(tr), Jack Sullivan/Alamy for p. 135(bl), Eurotech Ltd for p. 135(br), Nokia for p. 138(t), Jim Goldstein/Alamy for p. 139(br), Kim Kulish/Corbis for p. 145(t), Electronic Arts Limited for p. 145(tcr), Microsoft Xbox 360 box shot reprinted with permission from Microsoft Corporation for p. 145(tcl), Justin Leighton/Alamy for p. 145(cr), S.T. Yiap Conceptual/Alamy for p.145(br), Microsoft Game Studios for p. 146, © Honda for p. 150(tl), Don Farrall/Getty Images for p. 152, Cartoon by Patrick Blower for p. 154.

Logos and screenshots

p. 65 Microsoft Windows Vista Ultimate Desktop screenshot reprinted with permission from Microsoft Corporation; p. 84 screenshot http://www.cambridge.org/elt/resources/professional/, © Cambridge University Press 2005; p. 85 eBay screenshot. These materials have been reproduced with the permission of eBay Inc. © EBAY INC. All Rights Reserved. p. 85 Facebook screenshot, Facebook © 2007; p. 85 MySpace screenshot, © 2003-2007 MySpace. All Rights Reserved. p. 90 Microsoft Live Messenger logo, © 2007 Microsoft Corporation; p. 114: Yahoo! screenshot, reproduced with permission of Yahoo! Inc. © 2007 by Yahoo! Inc. YAHOO! and the YAHOO! logo are trademarks of Yahoo! Inc. p. 118 The TPS report screenshot, www.tpsreport.co.uk, Matthew Robinson © 2007; p. 125 Java Powered logo with permission of Sun Microsystems, Inc. Java and the Java Coffee Cup logo are trademarks or registered trademarks of Sun Microsystems, Inc. in the United States and other countries and are used by permission. p. 145: World of Warcraft screenshot, © 2007 Blizzard Entertainment Inc. All Rights Reserved. All Microsoft product screenshots reprinted with permission from Microsoft Corporation.

Designed and produced by eMC Design Ltd, www.emcdesign.org.uk

Picture research by Veena Holkar. Audio production by Penelope Reid.

1 Computers today

Learning objectives

In this module, you will:

- talk and write about computer applications in everyday life.
- study the basic structure of a computer system.
- study the differences between certain types of computer.
- learn how to classify computer devices.
- learn about the structure and functions of the CPU.
- learn how to distinguish between RAM and ROM.
- learn about how memory is measured.
- learn and use relative pronouns.
- learn how to enquire about computers in a shop.
- learn how to understand the technical specs of different computers.

Unit 1

Living in a digital age

1 *The digital age*

A Match the captions (1–4) with the pictures (a–d).

1 In education, computers can make all the difference ....
2 Using a cashpoint, or ATM ....
3 The Internet in your pocket ....
4 Controlling air traffic ....

a

b

c

d

B How are computers used in the situations above? In pairs, discuss your ideas.

C Read the text and check your answers to B.

The digital age

We are now living in what some people call *the digital age*, meaning that computers have become an essential part of our lives. Young people who have grown up with PCs and mobile phones are often called *the digital generation*. Computers help students to **perform** mathematical **operations** and improve their maths skills. They are used to **access the Internet**, to **do** basic **research** and to communicate with other students around the world. Teachers use projectors and interactive whiteboards to **give presentations** and teach sciences, history or language courses. PCs are also used for administrative purposes – schools use word processors to **write letters**, and databases to **keep records** of students and teachers. A school website allows teachers to publish **exercises** for students to **complete** online.

Students can also enrol for courses via the website and parents can download official reports.

Mobiles let you **make** voice **calls**, **send texts**, email people and download logos, ringtones or games. With a built-in camera you can send pictures and make video calls in *face-to-face* mode. New smartphones combine a telephone with web access, video, a games console, an MP3 player, a personal digital assistant (PDA) and a GPS navigation system, all in one.

In banks, computers **store information** about the money held by each customer and enable staff to **access** large **databases** and to **carry out** financial **transactions** at high speed. They also control the cashpoints, or ATMs (automatic teller machines), which **dispense money** to customers by the use of a PIN-protected card. People use a Chip and PIN card to pay for goods and services. Instead of using a signature to verify payments, customers are asked to **enter a** four-digit **personal identification number** (**PIN**), the same number used at cashpoints; this system makes transactions more secure. With online banking, clients can easily **pay bills** and **transfer money** from the comfort of their homes.

Airline pilots use computers to help them control the plane. For example, monitors **display data** about fuel consumption and weather conditions. In airport control towers, computers are used to manage radar systems and regulate air traffic. On the ground, airlines are connected to travel agencies by computer. Travel agents use computers to find out about the availability of flights, prices, times, stopovers and many other details.

D When you read a text, you will often see a new word that you don't recognize. If you can identify what type of word it is (noun, verb, adjective, etc.) it can help you guess the meaning.

Find the words (1–10) in the text above. Can you guess the meaning from context? Are they nouns, verbs, adjectives or adverbs? Write *n*, *v*, *adj* or *adv* next to each word.

1 perform (line 6) ....
2 word processor (line 13) ....
3 online (line 16) ....
4 download (line 18) ....
5 digital (line 25) ....
6 built-in (line 21) ....
7 store (line 27) ....
8 financial (line 29) ....
9 monitor (line 42) ....
10 data (line 42) ....

E Match the words in D (1–10) with the correct meanings (a–j).

a keep, save ....
b execute, do ....
c monetary ....
d screen ....
e integrated ....
f connected to the Internet ....
g collection of facts or figures ....
h describes information that is recorded or broadcast using computers ....
i program used for text manipulation ....
j copy files from a server to your PC or mobile ....

F In pairs, discuss these questions.

1 How are/were computers used in your school?
2 How do you think computers will be used in school in the future?

2 Language work: collocations 1

A Look at the HELP box and then match the verbs (1–5) with the nouns (a–e) to make collocations from the text on pages 2–3.

1	give	**a**	money
2	keep	**b**	a PIN
3	access	**c**	databases
4	enter	**d**	presentations
5	transfer	**e**	records

HELP box

Collocations 1

Verbs and nouns often go together in English to make set phrases, for example ***access the Internet***. These word combinations are called **collocations**, and they are very common. Learning collocations instead of individual words can help you remember which verb to use with which noun. Here are some examples from the text on pages 2–3: ***perform operations, do research, make calls, send texts, display data, write letters, store information, complete exercises, carry out transactions.***

B Use collocations from A and the HELP box to complete these sentences.

1. Thanks to Wi-Fi, it's now easy to .... from cafés, hotels, parks and many other public places.
2. Online banking lets you .... between your accounts easily and securely.
3. *Skype* is a technology that enables users to .... over the Internet for free.
4. In many universities, students are encouraged to .... using PowerPoint in order to make their talks more visually attractive.
5. The Web has revolutionized the way people .... – with sites such as *Google* and *Wikipedia*, you can find the information you need in seconds.
6. *Cookies* allow a website to .... on a user's machine and later retrieve it; when you visit the website again, it remembers your preferences.
7. With the latest mobile phones, you can .... with multimedia attachments – pictures, audio, even video.

3 Computers at work

A Listen to four people talking about how they use computers at work. Write each speaker's job in the table.

electrical engineer secretary librarian composer

Speaker	Job	What they use computers for
1		
2		
3		
4		

B Listen again and write what each speaker uses their computer for.

4 The magic of computers

A You are going to read a text about some of the other things that computers are used for. Five sentences have been removed from the text. Choose which sentence (1–5) fits which gap in the text (a–e).

1 It is a calculating machine that speeds up financial calculations
2 we visit shops and offices which have been designed with the help of computers
3 you can even use your PC to relax with computer games
4 for example calculators, the car's electronic ignition, the timer in the microwave, or the programmer inside the VCR
5 as does making a flight reservation or bank transaction

The magic of computers

Computers and microchips have become part of our everyday lives: (a) .... ; we pay bills prepared by computers; just picking up a telephone and dialling a number involves the use of a sophisticated computer system, (b) ....

Every day we encounter computers that spring to life the instant they are switched on, (c) ...., all of which use chip technology.

What makes your computer such a miraculous device? Each time you turn it on, it is a blank slate (*tabula rasa*) that, with appropriate hardware and software, is capable of doing anything you ask. (d) .... ; it is an electronic filing cabinet which manages large collections of data, such as customers' lists, accounts, or inventories; it is a magical typewriter that allows you to type and print any kind of document – letters, memos or legal documents; it is a personal communicator that enables you to interact with other computers and with people around the world; if you like gadgets and electronic entertainment, (e) ....

Nowadays, it is almost impossible to imagine life without the magic of computers.

B Read the text again and answer these questions.

1 Apart from computers, what other devices use microchips?
2 Which two components allow computer systems to operate?
3 What types of document are prepared on computers?
4 Why is a computer called a *personal communicator*?

Computers have changed the way we live, work, play and communicate

5 *Other applications*

A **In small groups, choose one of the areas in the diagram below and discuss what you can do with computers in that area. Look at the *Useful language* box below to help you.**

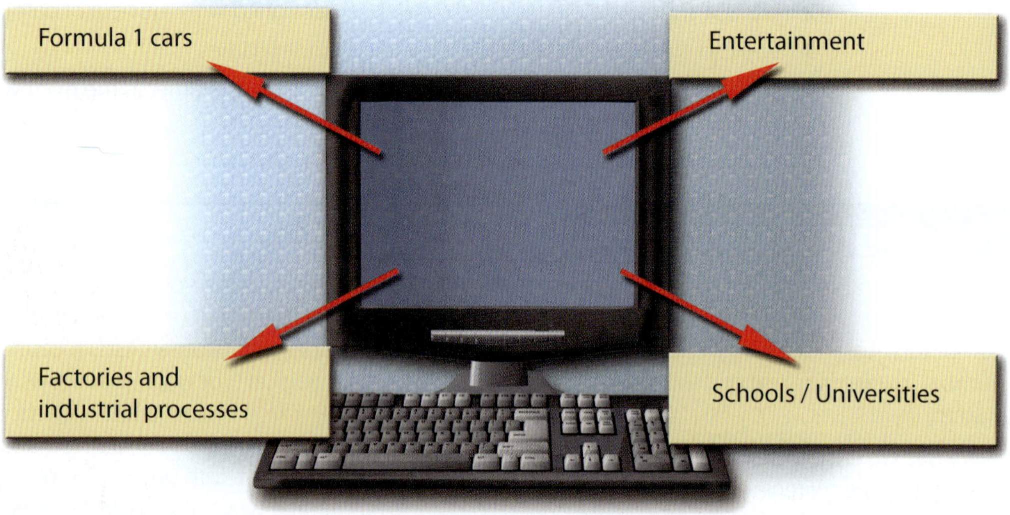

Useful language

Formula 1 cars: *design and build the car, test virtual models, control electronic components, monitor engine speed, store (vital) information, display data, analyse and communicate data*

Entertainment: *download music, burn CDs, play games, take photos, edit photos, make video clips, watch movies on a DVD player, watch TV on the computer, listen to MP3s, listen to the radio via the Web*

Factories and industrial processes: *design products, do calculations, control industrial robots, control assembly lines, keep record of stocks (materials and equipment)*

School/University: *access the Internet, enrol online, search the Web, prepare exams, write documents, complete exercises online, do research, prepare presentations*

Computers are used to …

A PC can also be used for …

People use computers to …

B **Write a short presentation summarizing your discussion. Then ask one person from your group to give a summary of the group's ideas to the rest of the class.**

Unit 2 Computer essentials

1 Computer hardware

A **In pairs, discuss these questions.**

1 Have you got a computer at home, school or work? What kind is it?
2 How often do you use it? What do you use it for?
3 What are the main components and features of your computer system?

B **In pairs, label the elements of this computer system.**

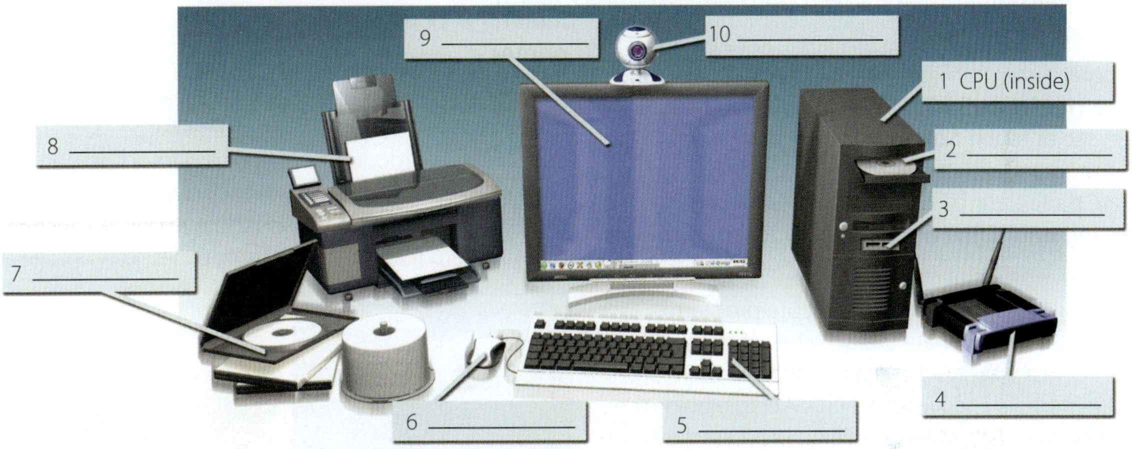

C **Read these advertising slogans and say which computer element each pair refers to.**

1
a Point and click here for power
b Obeys every impulse as if it were an extension of your hand

2
a Displays your ideas with perfect brilliance
b See the difference – sharp images and a fantastic range of colours

3
a It's quiet and fast
b *... it's easy to back up your data before it's too late*

4
a Power and speed on the inside
b Let your computer's brain do the work

5
a ... a big impact on the production of text and graphics
b Just what you need: a laser powerhouse

D **Find words in the slogans with the following meanings.**

1 to press the mouse button ....
2 clear; easy to see ....
3 to make an extra copy of something ....
4 selection ....
5 shows ....

2 What is a computer?

A Read the text and then explain Fig. 1 in your own words.

What is a computer?

A computer is an electronic machine which can accept data in a certain form, process the data, and give the results of the processing in a specified format as information.

First, data is fed into the computer's memory. Then, when the program is run, the computer performs a set of instructions and processes the data. Finally, we can see the results (the output) on the screen or in printed form (see Fig. 1 below).

A computer system consists of two parts: hardware and software. **Hardware** is any electronic or mechanical part you can see or touch. **Software** is a set of instructions, called a program, which tells the computer what to do. There are three basic hardware sections: the **central processing unit** (**CPU**), **main memory** and **peripherals**.

Perhaps the most influential component is the central processing unit. Its function is to execute program instructions and coordinate the activities of all the other units. In a way, it is the 'brain' of the computer. The main memory (a collection of RAM chips) holds the instructions and data which are being processed by the CPU. Peripherals are the physical units attached to the computer. They include storage devices and input/output devices.

Storage devices (hard drives, DVD drives or flash drives) provide a permanent storage of both data and programs. **Disk drives** are used to read and write data on disks. **Input devices** enable data to go into the computer's memory. The most common input devices are the **mouse** and the **keyboard**. **Output devices** enable us to extract the finished product from the system. For example, the computer shows the output on the **monitor** or prints the results onto paper by means of a **printer**.

On the rear panel of the computer there are several **ports** into which we can plug a wide range of peripherals – a modem, a digital camera, a scanner, etc. They allow communication between the computer and the devices. Modern desktop PCs have USB ports and memory card readers on the front panel.

A USB port

A USB connector

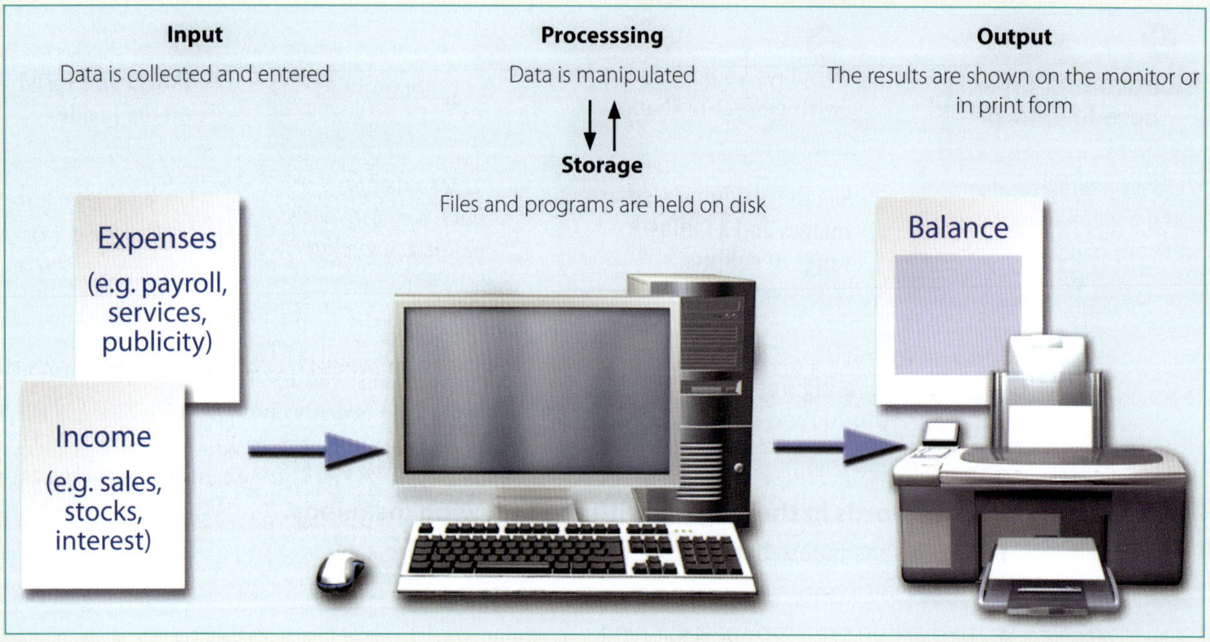

Fig. 1

B Match these words from the text (1–9) with the correct meanings (a–i).

1 software
2 peripherals
3 main memory
4 hard drive (also known as hard disk)
5 hardware
6 input
7 ports
8 output
9 central processing unit (CPU)

a the brain of the computer
b physical parts that make up a computer system
c programs which can be used on a particular computer system
d the information which is presented to the computer
e results produced by a computer
f input devices attached to the CPU
g section that holds programs and data while they are executed or processed
h magnetic device used to store information
i sockets into which an external device may be connected

3 *Different types of computer*

A Listen to an extract from an ICT class. As you listen, label the pictures (a–e) with words from the box.

laptop	desktop PC	PDA	mainframe	tablet PC

a ....

b ....

c ....

d ....

e ....

B Listen again and decide whether these sentences are true or false. Correct the false ones.

1 A mainframe computer is less powerful than a PC.
2 A mainframe is used by large organizations that need to process enormous amounts of data.
3 The most suitable computers for home use are desktop PCs.
4 A laptop is not portable.
5 Laptops are not as powerful as desktop PCs.
6 Using a stylus, you can write directly onto the screen of a tablet PC.
7 A Personal Digital Assistant is small enough to fit into the palm of your hand.
8 A PDA does not allow you to surf the Web.

4 *Language work: classifying*

A Look at the HELP box and then use suitable classifying expressions to complete these sentences.

1 A computer ______________ hardware and software.

2 Peripherals ______________ three types: input, output and storage devices.

3 A word processing program ______________ software which lets the user create and edit text.

4 ______________ of network architecture: peer-to-peer, where all computers have the same capabilities, and client-server (e.g. the Internet), where servers store and distribute data, and clients access this data.

B In pairs, describe this diagram, using classifying expressions from the HELP box. Make reference to your own devices.

HELP box

Classifying

Classifying means putting things into groups or classes. We can classify types of computers, parts of a PC, etc. Some typical expressions for classifying are:

- **... are classified into X types/categories**
- **... are classified by ...**
- **... can be divided into X types/categories**

  Digital computers can be ***divided into*** *five main* ***types****: mainframes, desktop PCs, laptops, tablet PCs and handheld PDAs.*
- **... include(s) ...**
- **... consist(s) of ...**

  The basic configuration of a mainframe ***consists of*** *a central system which processes immense amounts of data very quickly.*
- **There are X types/classes of ...**
- **X is a type of ...**

  A tablet PC is ***a type of*** *notebook computer.*

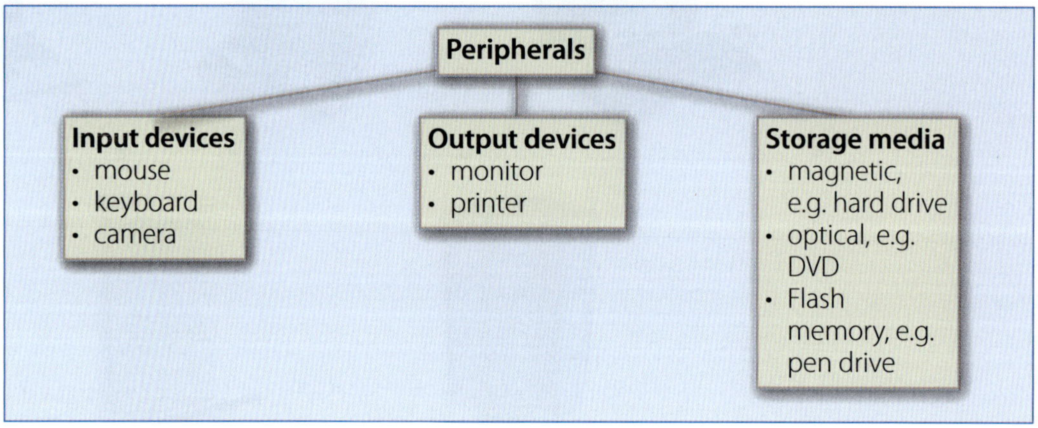

5 *Benefits of laptops and tablet PCs*

Your school is considering buying tablet PCs to use in the classroom. Write an email to your teacher explaining the benefits for the students and the school.

or

Your company is considering replacing all of the office PCs with laptops. Write an email to your boss explaining the benefits for the employees and the company.

Unit 3 Inside the system

1 *Technical specifications*

A Read the advertisement and translate the technical specifications into your own language.

Dell Inspiron 9200

- Intel Core 2 Duo processor at 2.4GHz
- 2048MB RAM, expandable to 4GB
- 500GB hard drive
- Comes with Windows Vista Home Premium

B In pairs, answer these questions. If necessary, look at the Glossary.

1. What is the main function of a computer's processor?
2. What unit of frequency is used to measure processor speed?
3. What does RAM stand for?

2 *What is inside a PC system?*

A Read the text on page 12 and then answer these questions.

1. What are the main parts of the CPU?
2. What does ALU stand for? What does it do?
3. What is the function of the system clock?
4. How much is one gigahertz?
5. What type of memory is temporary?
6. What type of memory is permanent and includes instructions needed by the CPU?
7. How can RAM be increased?
8. What term is used to refer to the main printed circuit board?
9. What is a *bus*?
10. What is the benefit of having expansion slots?

B Look at these extracts from the text. What do the words in bold refer to?

1. **This** is built into a single chip. (line 2)
2. … **which** executes program instructions and coordinates … (line 3)
3. … **that** is being executed. (line 22)
4. … performance of a computer is partly determined by the speed of **its** processor. (line 25)
5. … the CPU looks for **it** on the hard disk … (line 35)
6. … inside the computer to communicate with **each other**. (line 52)

What is inside a PC system?

Processing

The nerve centre of a PC is the **processor**, also called the **CPU**, or **central processing unit**. This is built into a single **chip** which executes program instructions and coordinates the activities that take place within the computer system. The chip itself is a small piece of silicon with a complex electrical circuit called an **integrated circuit**.

The processor consists of three main parts:

- The **control unit** examines the instructions in the user's program, interprets each instruction and causes the circuits and the rest of the components – monitor, disk drives, etc. – to execute the functions specified.
- The **arithmetic logic unit** (**ALU**) performs mathematical calculations (+, -, etc.) and logical operations (AND, OR, NOT).
- The **registers** are high-speed units of memory used to store and control data. One of the registers (the program counter, or PC) keeps track of the next instruction to be performed in the main memory. The other (the instruction register, or IR) holds the instruction that is being executed (see Fig. 1 on page 13).

The power and performance of a computer is partly determined by the speed of its processor. A **system clock** sends out signals at fixed intervals to measure and synchronize the flow of data. **Clock speed** is measured in **gigahertz** (**GHz**). For example, a CPU running at 4GHz (four thousand million hertz, or cycles, per second) will enable your PC to handle the most demanding applications.

The Intel Core 2 Duo processor; other chip manufacturers are AMD and Motorola

RAM and ROM

The programs and data which pass through the processor must be loaded into the main memory in order to be processed. Therefore, when the user runs a program, the CPU looks for it on the hard disk and transfers a copy into the **RAM** chips. RAM (**random access memory**) is volatile – that is, its information is lost when the computer is turned off. However, **ROM** (**read only memory**) is non-volatile, containing instructions and routines for the basic operations of the CPU. The **BIOS** (**basic input/output system**) uses ROM to control communication with peripherals.

RAM capacity can be expanded by adding extra chips, usually contained in small circuit boards called dual in-line memory modules (**DIMMs**).

A RAM chip

Buses and cards

The main circuit board inside your system is called the **motherboard** and contains the processor, the memory chips, expansions slots, and controllers for peripherals, connected by **buses** – electrical channels which allow devices inside the computer to communicate with each other. For example, the front side bus carries all data that passes from the CPU to other devices.

The size of a bus, called **bus width**, determines how much data can be transmitted. It can be compared to the number of lanes on a motorway – the larger the width, the more data can travel along the bus. For example, a 64-bit bus can transmit 64 bits of data.

Expansion slots allow users to install **expansion cards**, adding features like sound, memory and network capabilities.

A data bus

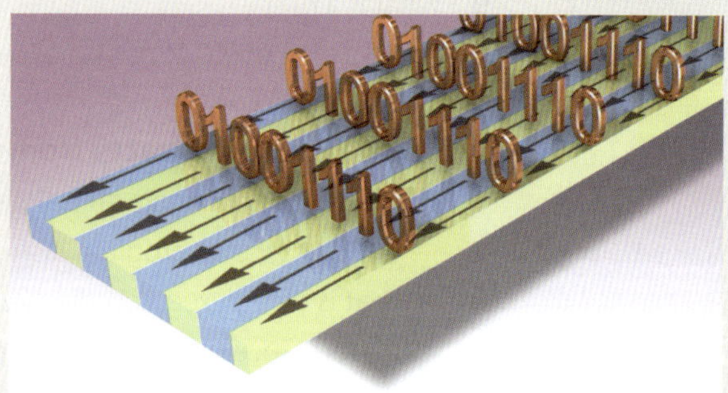

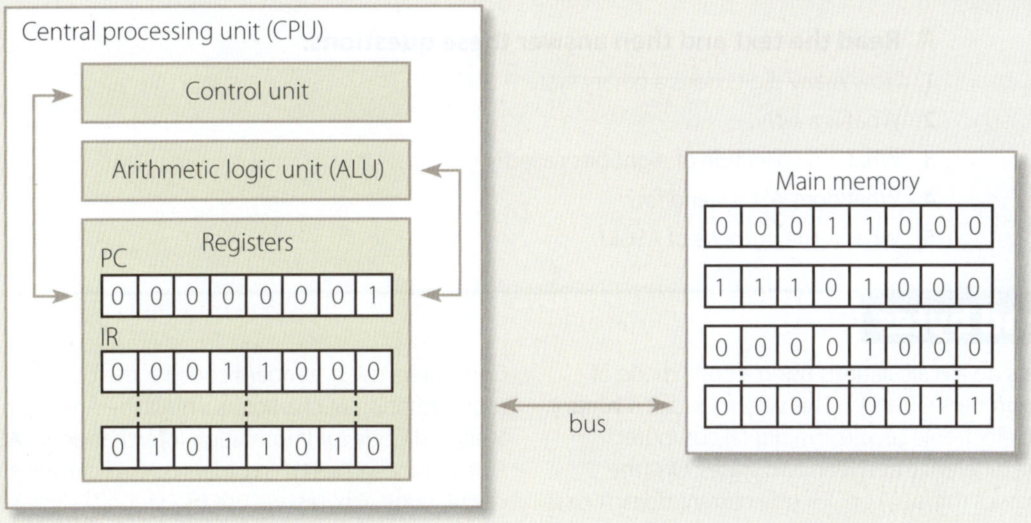

Fig. 1: Organization of a simple computer; the CPU is built into a single microprocessor chip

3 Language work: defining relative clauses

Look at the HELP box and then complete the sentences below with suitable relative pronouns. Give alternative options if possible. Put brackets round the relative pronouns you can leave out.

1 That's the computer .... I'd like to buy.

2 Core 2 Duo is a new Intel processor .... contains about 291 million transistors.

3 A webmaster is a person .... designs, develops and maintains a website.

4 A bus is an electronic pathway .... carries signals between computer devices.

5 Here's the DVD .... you lent me!

6 Last night I met someone .... works for GM as a software engineer.

HELP box

Defining relative clauses

- We can define people or things with a defining (restrictive) relative clause. We use the relative pronoun **who** to refer to a person; we can also use **that**.

  A blogger is a person ***who/that*** *keeps a web log (blog) or publishes an online diary.*

- We use the relative pronoun **which** (or **that**) to refer to a thing, not a person.

  This is built into a single chip ***which/that*** *executes program instructions and coordinates the activities that take place within the computer system.*

- Relative pronouns can be left out when they are the object of the relative clause.

  *The main circuit board (****which/that****) you have inside your system is called the motherboard …*

4 How memory is measured

A Read the text and then answer these questions.

1 How many digits does a binary system use?
2 What is a *bit*?
3 What is a collection of eight bits called?
4 What does ASCII stand for?
5 What is the purpose of ASCII?

Bits and bytes

Computers do all calculations using a code made of just two numbers – 0 and 1. This system is called **binary code**. The electronic circuits in a digital computer detect the difference between two states: ON (the current passes through) or OFF (the current doesn't pass through) and represent these states as 1 or 0. Each 1 or 0 is called a **binary digit**, or **bit**.

Bits are grouped into eight-digit codes that typically represent characters (letters, numbers and symbols). Eight bits together are called a **byte**. Thus, each character on a keyboard has its own arrangement of eight bits. For example, 01000001 for the letter A, 01000010 for B, and 01000011 for C.

Computers use a standard code for the binary representation of characters. This is the American Standard Code for Information Interchange, or **ASCII** – pronounced /'æski/. In order to avoid complex calculations of bytes, we use bigger units such as kilobytes, megabytes and gigabytes.

We use these units to describe the RAM memory, the storage capacity of disks and the size of a program or document.

Note: **bit** is pronounced /bɪt/; **byte** is pronounced /baɪt/

One bit

01000011

Example of a byte

Unit of memory	Abbreviation	Exact memory amount
Binary digit	bit, b	1 or 0
Byte	B	8 bits
Kilobyte	KB or K	1,024 bytes (2^{10})
Megabyte	MB	1,024 KB, or 1,048,576 bytes (2^{20})
Gigabyte	GB	1,024 MB, or 1,073,741,824 bytes (2^{30})
Terabyte	TB	1,024 GB, or 1,099,511,627,776 bytes (2^{40})

B Complete these descriptions with the correct unit of memory.

1 A .... is about one trillion bytes – about as much text as the books and magazines in a huge library.
2 A .... is about one million bytes – about as much text as a 300-page novel.
3 A .... is about one thousand bytes – equivalent to one sheet of A4.
4 A .... is about one billion bytes – about as much text as 1,000 books.
5 A .... can store a single character, such as the letter *h* or number *7*.

5 A PC system

A Complete this diagram of a PC system. Look at Units 1, 2 and 3 to help you.

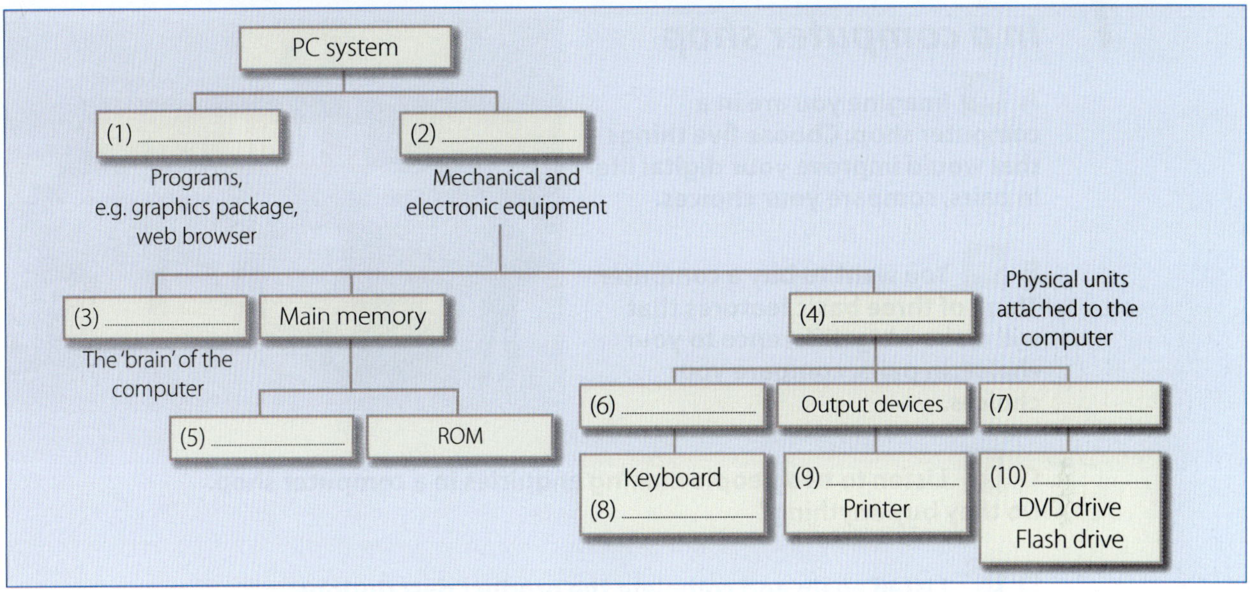

B In pairs, compare your answers.

C Listen to a teacher explaining the diagram to her class and check your answers.

6 Your ideal computer system

A Make notes about the features of the computer that you would most like to have. Think about the features in the box.

CPU   Speed   Optical disc drives   Wireless connectivity   Minimum/maximum RAM
Monitor   Ports and card memory slots   Hard disk   Software

B In pairs, describe your ideal computer system. Give reasons for your choices.

Useful language

It's got …

It's very fast. It runs at …

The standard RAM memory is … and it's expandable …

The hard disk can hold …

I need a large, flat LCD screen because …

As for the Internet, …

Unit 4 Buying a computer

1 *In a computer shop*

A **Imagine you are in a computer shop. Choose five things that would improve your digital life. In pairs, compare your choices.**

B **You want to buy a computer. Think of three basic features that will make a big difference to your choice. In pairs, compare your choices.**

C **Listen to two people making enquiries in a computer shop. Do they buy anything?**

D **Listen again and complete the product descriptions.**

iMac

Processor speed 2.33GHz

RAM ....

Hard drive capacity ....

DVD drive included? Yes

Operating system ....

Includes internet software

Price ....

MacBook

Processor speed ....

RAM ....

Hard drive capacity ....

DVD drive included? ....

Operating system ....

Includes internet software

Price £1,029

E **Listen again and complete the extract from the conversation.**

Assistant: Do you need any (1) .... ?

Paul: Um, yes, we're looking for a Mac computer. Have you got any fairly basic ones?

Assistant: Yes, sure. If you'd like to come over here.

Paul: What different (2) .... are there?

Assistant: At the moment we've got these two models: the iMac, which is a desktop computer with an Intel Core 2 Duo processor (3) .... at 2.33 gigahertz, and the portable MacBook, which has a processor (4) .... at 2.0 gigahertz. Core Duo technology actually means two cores, or processors, built into a single chip, offering up to twice the speed of a traditional chip.

Sue: So they're both very (5) .... , then. And which one has more memory? I mean, which has more RAM?

Assistant: Well, the iMac has two gigabytes of RAM, which can be (6) .... up to three gigabytes, and the MacBook has one gigabyte, expandable to two gigabytes. It all depends on your needs. The iMac is (7) .... for home users and small offices. The MacBook is more (8) .... if you travel a lot.

2 Language functions in a computer shop

Look at the language functions in the HELP box and then correct one mistake in each of these sentences. Decide which functions are being expressed in each sentence.

1 The Ulysses SD is a power, expandable computer that offers high-end graphics at a low price.
2 A laptop is likely to be more expensive than the equivalent desktop, but a laptop is less practical if you travel a lot.
3 Where's the storage capacity of the hard drive?
4 I'm looking a desktop PC that has good graphics for games.
5 Do you need the help?
6 And how many does the PDA cost?
7 This workstation is a Pentium processor with dual-core technology, 1,024 gigabytes of RAM, and 1 terabyte of disk space.

HELP box

Language functions useful to a sales assistant

- Greeting and offering help
  Good morning. Do you need any help?
- Giving technical specifications (specs)
  The MacBook has a processor running at 2.0 gigahertz.
  The iMac has two gigabytes of RAM.
  They feature a camera built into the display.
- Describing
  Both computers are very fast and reliable.
- Comparing
  The MacBook is more practical if you travel a lot.
  PDAs are cheaper than laptops but laptops are more powerful.

Language functions useful to a customer

- Explaining what you are looking for
  We're looking for a personal computer. Have you got any fairly basic ones?
- Asking for technical specs
  What's the storage capacity of the hard drive?
  Do they have a DVD drive?
- Asking the price
  How much do they cost?
  How much is it?

3 *Role play – buying a computer*

Work in pairs. One of you wants to buy a computer, the other is the shop assistant. Use the prompts and product descriptions below to role play the conversation.

Shop assistant	Customer
Greet the customer and offer help.	
	Explain what you are looking for.
Show the customer two possible models.	
	Ask for some technical specs.
Give technical specs (describe the processor, RAM and storage capacity). Compare the two different models.	
	Ask about any further technical specs (DVD drive, monitor, communications, etc.).
Give the information required. Compare the two models.	
	Ask the price.
Answer, and mention any final details that might persuade the customer to buy the computer.	
	Decide which computer to buy or leave the shop.

Toshiba Satellite laptop
2.0GHz Core 2 Duo processor
2GB RAM expandable to 4GB
160GB hard drive
Super Multi drive (double layer)
15.4" wide XGA display
Wireless LAN, Wi-Fi compliancy
£1,099

Dell desktop PC
AMD Athlon at 2.4GHz
1GB RAM expandable to 4GB
320GB hard drive
DVD+/-RW drive
17" LCD monitor
£680

Palm TX handheld
Intel 312MHz ARM-based processor
128 MB Flash memory (non-volatile)
Support for memory cards
320x480 TFT touch screen
Wi-Fi and Bluetooth
Lithium-ion battery
£216

4 *Choosing the right computer*

A Listen to four people talking about their computer needs and take notes. In pairs, read the descriptions from the computer shop website and choose the most suitable computer for each person. Give reasons for your choices.

Speaker 1 .... Speaker 3 ....

Speaker 2 .... Speaker 4 ....

Sun workstation
Two AMD Opteron processors at 3.0GHz
4GB RAM; 32GB maximum
1 terabyte hard drive and dual DVD drive
19" Sun TFT flat-panel LCD
Supports several graphics formats
Allows you to handle your toughest technical, scientific, and business-critical applications
Supports Solaris, Windows and Linux
£3,249

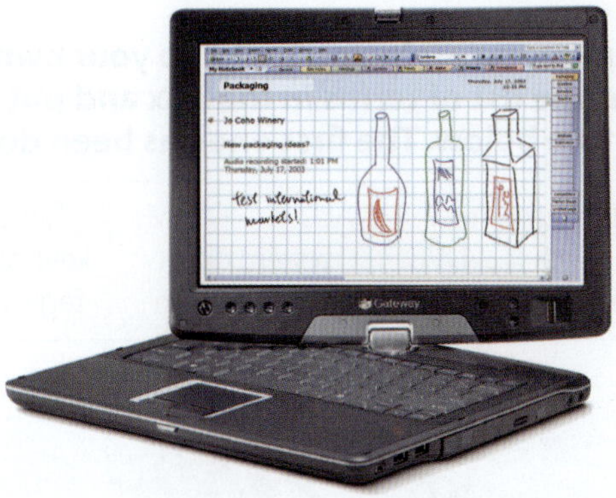

Gateway C-120 convertible notebook
Intel Core 2 Duo ULV processor at 1.06GHz
12.1" WXGA TFT touch screen
Gateway Executive stylus pen
1024MB DDR2 SDRAM
80GB serial ATA hard drive
DVD-ROM drive (optical DVD burner)
Integrated modem and Bluetooth
Windows Vista Home Premium
Thin and lightweight (1.17", 2.4 kg)
£805

Sony Vaio AR laptop (VGN-AR51E)
Intel Core 2 Duo Processor at 2GHz
2GB DDR2 SDRAM
200GB hard drive
DVD+/-RW optical drive
17" WXGA high-definition LCD screen
Memory Stick slot
Three USB 2.0 ports
Integrated wireless LAN
Built-in 'Motion Eye' digital camera
Lithium-ion battery
Windows Vista Ultimate
£899

Dell Inspiron 531 desktop PC
AMD Athlon 64 X2 Dual Core Processor
3072MB DDR2 SDRAM
Dell 22" Wide Flat Panel
256MB NVIDIA GeForce 8600GT video card
1.0TB Hard Drive
16x DVD+/- RW Drive
Integrated 7.1 Channel High Definition Audio
Windows Vista Home Premium
Optional features: Windows Media Center, integrated TV Tuner, and a Blu-ray disc drive for high-definiton content
From £849

B **Look at the notes you made about your ideal computer system in Unit 3 task 6 (page 15). What did you want? Look again at the descriptions of the computers above and choose the one that is closest to your ideal. In pairs, discuss your choices.**

5 Vocabulary tree

Designing word trees and spidergrams can help you build up your own mental 'maps' of vocabulary areas. Look at the list of terms in the box and put each one in an appropriate place on the word tree below. The first one has been done for you.

~~processor~~	ROM	expandable memory	ALU	DIMMs	hard drive
RAM	computer brain	byte	DVD	system clock	keyboard
mouse	gigahertz	printer	megabyte	webcam	registers

6 Recommending a computer

A friend has asked you to recommend a computer that suits his needs. He needs to be able to access the Internet, play games and work with graphics, music and video files. Write an email describing its technical features and saying why you recommend it.

Now visit www.cambridge.org/elt/ict for an online task.

2 Input/Output devices

Learning objectives

In this module, you will:

- describe input and output devices.
- identify the different keys on a keyboard and explain their functions.
- distinguish between facts and opinions in advertisements.
- learn how to understand the technical specs of digital cameras, printers and display devices.
- learn and use the superlative form of adjectives.
- practise recommending the most suitable display device for particular people.
- learn how to understand and give instructions and advice for the use of computers and monitors.
- compare different types of printer.
- learn and use discourse connectors.
- learn about what sort of input/output devices disabled people can use.

Unit 5 Type, click and talk!

1 *Interacting with your computer*

Read the description of input devices and then label the pictures (1–8) with words from the text.

Input devices are the pieces of hardware which allow us to enter information into the computer. The most common are the **keyboard** and the **mouse**. We can also interact with a computer by using one of these: a **light pen**, a **scanner**, a **trackball**, a **graphics tablet**, a **game controller** or a **microphone**.

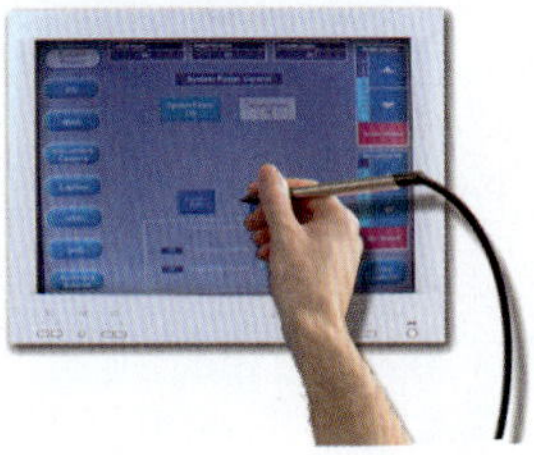

1 ....

2 ....

3 ....

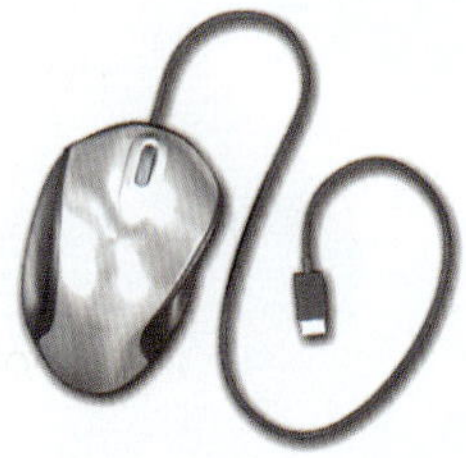

4 ....

5 ....

6 ....

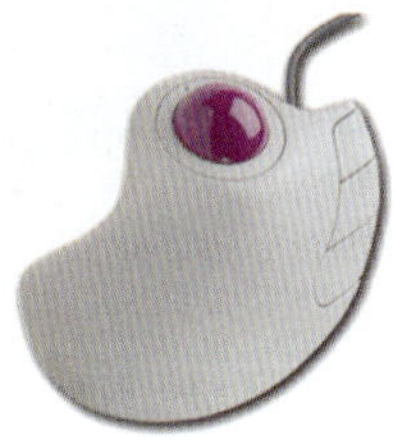

7 ....

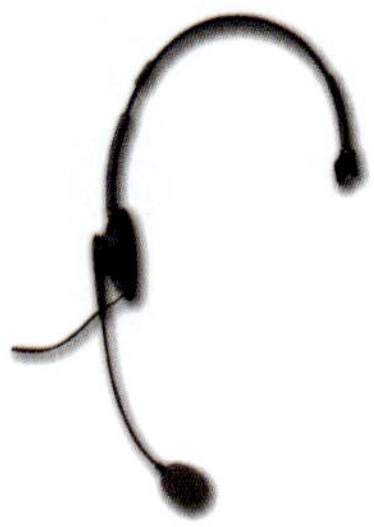

8 ....

2 *Describing input devices*

A Listen to a computer technician describing three input devices. Write which devices he's talking about.

1 .... **2** .... **3** ....

B Listen again and complete these extracts.

1 This device is .... enter information into the computer.

2 ... it may also .... function keys and editing keys .... special purposes.

3 This is a device .... the cursor and selecting items on the screen.

4 It usually .... two buttons and a wheel.

5 ... the user .... activate icons or select items and text.

6 It .... detecting light from the computer screen and is used by pointing it directly at the screen display.

7 It .... the user .... answer multiple-choice questions and ...

3 Describing functions and features

A Look at the HELP box and then use the notes below to write a description of the Sony PlayStation 3 controller.

Sony PlayStation 3 controller

Functions

- control video games
- hold it with both hands, use thumbs to handle directional sticks and face buttons

Features

- six-axis sensing system (capable of sensing motion in six directions: up, down, left, right, forwards and backwards)
- wireless controller (Bluetooth)
- USB mini port and cable for wired play and automatic battery charging

HELP box

Describing functions

In the listening, the mouse was described using **for** + gerund:

This is a device ***for controlling*** *the cursor and selecting items on the screen.*

There are other ways of describing a device's function:

- **used** + **to** + infinitive
  It's ***used to control*** *...*
- relative pronoun + verb
  This is a device ***which controls*** *...*
- relative pronoun + **used** + **to** + infinitive
  This is a device ***which/that*** *is* ***used to control*** *...*
- **work by** + gerund
  It ***works by detecting*** *light from the computer screen.*

Describing features

We can describe features like this:

An optical mouse ***has*** *an optical sensor instead of a ball underneath.*

It usually ***features*** *two buttons and a wheel.*

You ***can*** *connect it to a USB port.*

A wireless mouse ***works/operates*** *without cables.*

It ***allows*** *the user* ***to*** *answer multiple-choice questions and ...*

B In pairs, choose one of these input devices and describe its functions and features. Try to guess which device your partner is describing.

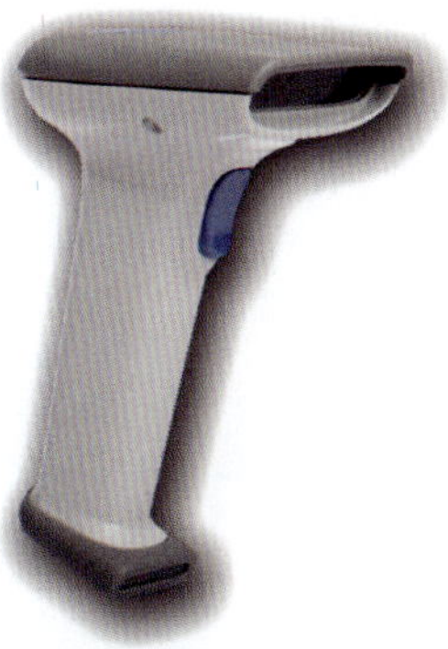

Bar code reader

Touchpad on a portable PC

Webcam

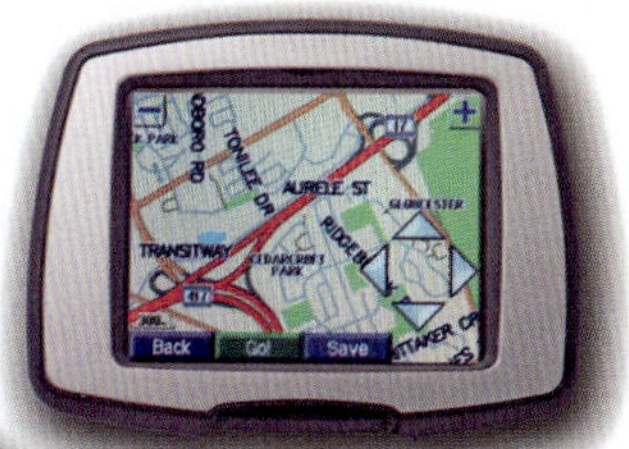

Touch screen

4 The keyboard

A Label the picture of a standard keyboard with the groups of keys (1–5).

1 **Cursor control keys** include arrow keys that move the insertion point up, down, right and left, and keys such as *End*, *Home*, *Page Up* and *Page Down*, which are used in word processing to move around a long document.

2 **Alphanumeric keys** represent letters and numbers, as arranged on a typewriter.

3 **Function keys** appear at the top of the keyboard and can be programmed to do special tasks.

4 **Dedicated keys** are used to issue commands or to produce alternative characters, e.g. the *Ctrl* key or the *Alt* key.

5 A **numeric keypad** appears to the right of the main keyboard. The *Num Lock* key is used to switch from numbers to editing keys.

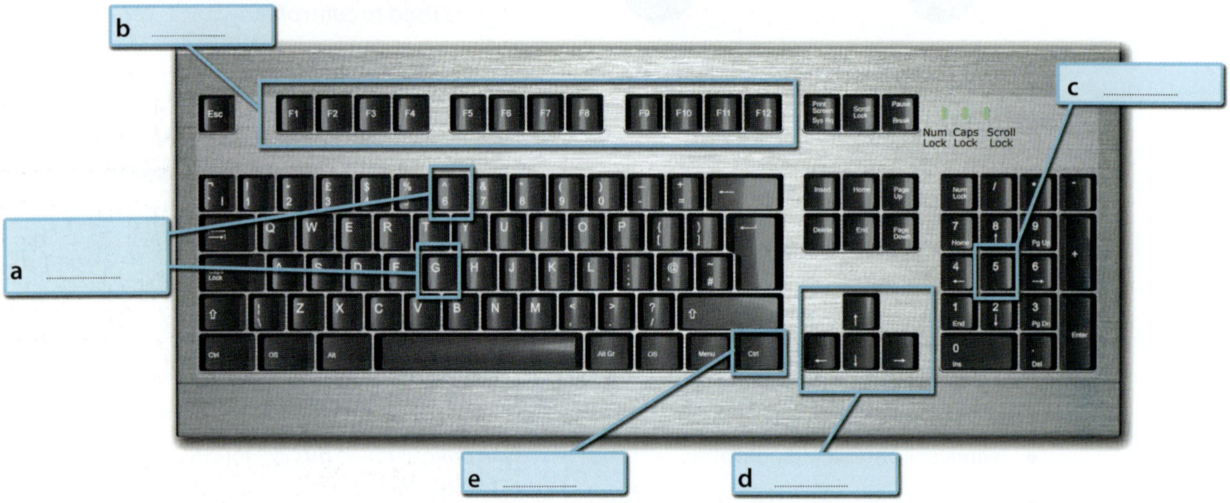

A PC-compatible keyboard

B Match the descriptions (1–8) with the names of the keys (a–h). Then find them on the keyboard.

1 A long key at the bottom of the keyboard. Each time it is pressed, it produces a blank space.

2 It moves the cursor to the beginning of a new line. It is also used to confirm commands.

3 It works in combination with other keys. For example, you press this key and *C* to copy the selected text.

4 It removes the character to the left of the cursor or any selected text.

5 It produces UPPER CASE characters.

6 It produces UPPER CASE letters, but it does not affect numbers and symbols.

7 It moves the cursor horizontally to the right for a fixed number of spaces (in tabulations and data fields).

8 They are used to move the cursor, as an alternative to the mouse.

a arrow keys
b return/enter
c Caps Lock
d shift
e tab
f space bar
g backspace
h Ctrl

5 Mouse actions

Complete this text about the mouse with verbs from the box.

click	double-click	drag	grab	select	move	control

Mouse actions

A mouse allows you to (1) .... the cursor and move around the screen very quickly. Making the same movements with the arrow keys on the keyboard would take much longer. As you (2) .... the mouse on your desk, the pointer on the screen moves in the same direction. The pointer usually looks like an I-bar, an arrow, or a pointing hand, depending on what you are doing.

A mouse has one or more buttons to communicate with the computer. For example, if you want to place the insertion point or choose a menu option, you just (3) .... (press and release) on the mouse button, and the option is chosen.

The mouse is also used to (4) .... text and items on the screen. You can highlight text to be deleted, copied or edited in some way.

The mouse is widely used in graphics and design. When you want to move an image, you position the pointer on the object you want to move, press the mouse button, and (5) .... the image to a new location on the screen. Similarly, the mouse is used to change the shape of a graphic object. For example, if you want to convert a square into a rectangle, you (6) .... one corner of the square and stretch it into a rectangle.

The mouse is also used to start a program or open a document: you put the pointer on the file name and (7) .... on the name – that is, you rapidly press and release the mouse button twice.

www.CartoonStock.com

6 *Speech recognition systems*

A Listen to an interview with Anne Simpson, an expert in voice input technologies and tick (✔) the features she mentions.

Speech recognition systems:

- ☐ need a good sound card and a microphone.
- ☐ can take dictation with accuracy.
- ☐ allow you to create and compile a computer program.
- ☐ allow you to execute programs and navigate around menus using voice commands.
- ☐ allow you to surf the Web by speaking.
- ☐ allow you to design graphics.

B Listen again and answer these questions.

1 What do people usually use to communicate with a computer?
2 How do you get the best results from speech recognition software?
3 What rate of accuracy is possible with the software?
4 How can you train the software to be more accurate?
5 What kinds of words aren't in the software's dictionary?

C In groups, discuss these questions.

1 What are the benefits of speech recognition software?
2 What kind of tasks would you find speech recognition useful for?
3 Who would benefit most from advances in speech recognition technology?
4 What is the future of this kind of technology? Do you think it will ever be possible to control your computer using only your thoughts?

You talk, it types – speech recognition software lets you operate computers by voice command

Unit 6 Capture your favourite image

1 *The eyes of your computer*

A In pairs, discuss how many ways there are of capturing an image on a computer.

B Read the text and see how many things from your list are mentioned.

C Read the text again and answer these questions.

1 Which device is used to input text and graphic images from a printed page?

2 How does a scanner send information to the computer?

3 How do digital cameras store photographs?

4 What feature allows mobile phone users to take pictures?

5 Which device would you use to take digital video?

6 What kind of software is used to manipulate video clips on the computer?

The eyes of your computer

What does a scanner do?

A scanner 'sees' images and converts the printed text or pictures into electronic codes that can be understood by the computer. With a flatbed colour scanner, the paper with the image is placed face down on a glass screen, as with a photocopier. Beneath the glass are the lighting and measurement devices. Once the scanner is activated, it reads the image as a series of dots and then generates the digitized image that is sent to the computer and stored as a file.

The scanner operates by using three rotating lamps, each of which has a different coloured filter: red, green and blue. The resulting three separate images are combined into one by appropriate software.

What does a digital camera do?

A digital camera takes photos electronically and converts them into digital data (binary codes made up of 1s and 0s). It doesn't use the film found in a traditional camera; instead it has a special light-sensitive silicon chip.

Photographs are stored in the camera's memory card before being sent to the computer. Some cameras can also be connected to a printer or a TV set to make viewing images easier. This is usually the case with camera phones – mobile phones with a built-in camera.

What does a camcorder do?

A camcorder, or digital video camera, records moving pictures and converts them into digital data that can be stored and edited by a computer with special video editing software.

Digital video cameras are used by home users to create their own movies, or by professionals in computer art and video conferencing.

They are also used to send live video images via the Internet. In this case they are called web cameras, or webcams.

2 Scanners

Listen to a conversation between Vicky Cameron, an Information Technology (IT) lecturer, and one of her students, and complete the student's notes.

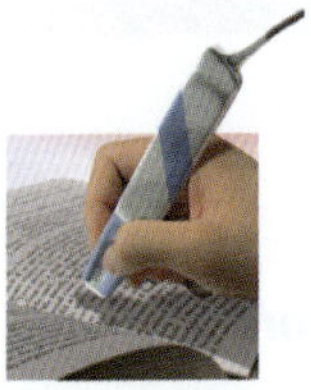

A handheld scanner for scanning text, bar codes and handwritten numbers

1 The technology used in scanners is similar to that used in a ....
2 The scanned image is sent to the ...., where you can manipulate it.
3 To scan text, you need special software called ....
4 Flatbed scanners can scan ....
5 Slide scanners are used to scan .... or film negatives.
6 Handheld scanners are used for capturing ....

A slide scanner

3 Facts and opinions

A What is the difference between facts and opinions? Complete these definitions.

1 .... are real, objective information.
2 .... usually include emotive words and subjective statements.

B Read these advertisements and underline the facts and circle the opinions.

ColourScan XR
from Sunrise

The ColourScan XR from Sunrise is a flatbed scanner with 1,200 dots per inch (dpi) of resolution and 9"x15" of scanning area.

Just think of the possibilities.

You can enter data and graphic images directly into your applications (word processors or databases). You can get crisp, clean scans for colour compositions, video and animation work.

The ColourScan XR comes complete with its own image-capture software, which allows for colour and grey retouching. *And* it's easy to use. What more could you want for only £79? It couldn't be cheaper.

In the field of flatbeds, the ColourScan XR is the clear winner.

ScanPress DF

The ScanPress DF is a self-calibrating flatbed scanner with 2,400 dpi of resolution. You can scan everything from black and white to 24-bit colour. The package includes a hardware accelerator for JPEG compression and decompression. JPEG technology saves disk space by compressing images by up to 50 to 1.

In creating the ScanPress DF, we have chosen the most advanced technology to give you the best scans with the least effort. It produces images with high colour definition and sharpness. And it comes with OCR software and Adobe Photoshop, so you can manipulate all the images you capture.

The ScanPress DF is a fantastic machine that you will love working with. And at only £309, an excellent investment.

C In small groups, compare your answers and decide about the following.

1 Which text uses more persuasive language?

2 Which text is more factual or objective?

4 *Language work: superlatives*

A Apart from catchy slogans and other persuasive techniques, advertisements often use the superlative form of adjectives and adverbs. Read the following examples from advertisements for input devices. What can you say from these examples about how superlatives are formed? Look at the HELP box to check your answers.

1 We have chosen the most advanced technology …
2 The fastest personal scanner …
3 The most revolutionary computer peripheral …
4 The best scans with the least effort …

B Complete these sentences with the superlative form of the adjectives in brackets.

1 Always buy the (fast) .... scanner with the (high) .... resolution you can afford.

2 They have created the (revolutionary) .... camera to date.

3 FotoFinish is the (easy) .... photo editing software for your digital camera.

4 This scanner gives you the (good) .... scans with the (little) .... effort.

5 Our university has bought the (modern) .... computer equipment.

HELP box

Superlatives

- We form the superlative of one-syllable and most two-syllable adjectives by adding **-est**.

  cheap → ***the** cheap**est***
  clever → ***the** clever**est***

- Some two-syllable adjectives (including those ending in **-ing**, **-ed**, **-ful** and **-less**) form the superlative with **the most/least**.

  advanced → ***the most** advanced*

- Adjectives with three or more syllables also take **the most/least**.

  fantastic → ***the most** fantastic*
  powerful → ***the least** powerful*

- But two syllable adjectives ending in **-y** (for example, **noisy**) take **-est** and the **y** changes to **i**.

  noisy → ***the** nois**iest***

- Note the irregular forms:

  good → ***the best***
  bad → ***the worst***
  little → ***the least***
  (with amounts, not size)

C In pairs, discuss who or what you think is:

1 the most difficult computer game you've ever played.
2 the most exciting film you've ever seen.
3 the funniest programme on TV.
4 the most dangerous computer virus.
5 the best blogger or webmaster on the Web.
6 the most popular web browser.

5 *Language work: suffixes*

A Look at the HELP box and then use suitable suffixes to make adjectives or nouns from these words. In some cases, you can make more than one word. Use a dictionary to help you.

1 colour
2 profession
3 photograph
4 wire
5 blur
6 innovate
7 underexpose

HELP box

Suffixes

- Suffixes change the class of the root word. For example, by adding the suffix **-al**, the noun **digit** is changed into the adjective **digital**. Suffixes can help you tell if a word is a noun, an adjective or a verb.

  Common adjectival suffixes are: **-y**, **-able**, **-ible**, **-ive**, **-al**, **-ed**, **-ful**, **-ic**, **-less**, **-ing**

  Common noun suffixes are: **-er**, **-or**, **-ion**, **-tion**, **-ation**, **-ment**, **-ness**, **-ity**, **-ant**, **- logy**, **-ing**, **-y**, **-ure**, **-sion**

- When using suffixes, always check in your dictionary to see if you need to change any other letters.

  scan* → *scanner (double **n**)

B Complete these sentences with the word in brackets and one of these noun suffixes: *-tion, -er, -ing, -logy, -ness*. Use a dictionary to help you.

1 Kodak is a (manufacture) .... of photographic and imaging equipment.
2 To avoid red eyes, use the camera's red eye (reduce) .... feature.
3 (Crop) .... a photograph means cutting out the parts of an image you don't need.
4 The (sharp) .... of a photograph is a combination of resolution and acutance – the ability to represent clear edges.
5 Digital (techno) .... is evolving so rapidly that some cameras have a resolution of 12 megapixels – that's 12 million pixels.

6 Press release: a digital camera

Complete the press release with words from the box.

colour	megapixels	shot	video	optical	brighter	reduction

Kodak has introduced the EasyShare M753 digital camera, with 7.0 (1) .... resolution, a huge 2.5-inch LCD screen, and a professional 3x (2) .... zoom lens. It is the first camera to incorporate proprietary Kodak Perfect Touch Technology. At the touch of a button, this innovative feature creates better, (3) .... pictures by bringing out detail in shadows without affecting lighter areas. It's ideal for underexposed pictures caused by shooting beyond the flash range or in adverse lighting conditions.

The M753 uses the exclusive Kodak Colour Science chip for phenomenal image quality with rich (4) .... and accurate skin tones. Seventeen programmed scene modes (e.g. party, fireworks, children) and five colour modes (high, low, natural, sepia, and black and white) help capture the best (5) .... with the least effort.

Other features include cropping, auto picture rotation, digital red-eye (6) .... , and blurry picture alert. For capturing more than just still pictures, the camera also features high-quality (VGA) (7) .... capture and playback.

7 Describing a camera

In pairs, describe your digital camera, webcam or video camera. Think about these questions.

- What do you use the device for?
- Why did you buy that particular make/model?
- What are your favourite functions?
- What improvements would you make to the device?

Unit 7 Display screens and ergonomics

1 Your computer screen

In pairs, discuss these questions.

1 What type of display do you have: a cathode ray tube or an LCD flat screen?
2 What size is the screen?
3 How can you change the picture using the controls?
4 Can you watch TV on your PC monitor?

An Apple Mac flat screen monitor

2 How screen displays work

A Complete these definitions with words from the box. Then read the text on page 33 and check your answers.

resolution	pixel	aspect ratio	colour depth	video adapter	plasma screen

1 .... – the smallest unit on a display screen or bitmapped image (usually a coloured dot)
2 .... – an expansion card that generates the video signal sent to a computer display
3 .... – the width of the screen in proportion to its height
4 .... – also called *gas discharge display*
5 .... – the number of pixels contained in a display, horizontally and vertically
6 .... – the number of bits used to hold a colour pixel; this determines the maximum number of colours that can be displayed

B Read the text again and answer these questions.

1 What do CRT and LCD stand for?
2 How is the screen size measured?
3 What technology is used by active-matrix LCDs?
4 Which unit of frequency is used to measure the brightness of a display?
5 What substance produces light and colour when hit by electrons in a CRT monitor?
6 What are the three advantages of OLED displays?

How screen displays work

Displays, often called **monitors** or **screens**, are the most-used output device on a computer. They provide instant feedback by showing you text and graphic images as you work or play.

Most desktop displays use **Liquid Crystal Display** (**LCD**) or **Cathode Ray Tube** (**CRT**) technology, while nearly all portable computing devices, such as laptops, incorporate LCDs. Because of their slimmer design and lower energy consumption, LCD monitors (also called **flat panel** or **flat screen** displays) are replacing CRTs.

Basic features

Resolution refers to the number of dots of colour, known as **pixels** (picture elements), contained in a display. It is expressed by identifying the number of pixels on the horizontal and vertical axes. A typical resolution is 1024x768.

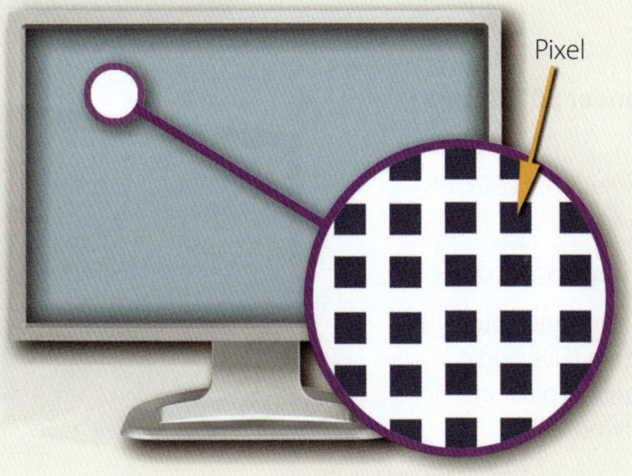

A pixel is a combination of red, green and blue subpixels

Two measurements describe the size of your display: the **aspect ratio** and the **screen size**. Historically, computer displays, like most televisions, have had an aspect ratio of 4:3 – the width of the screen to the height is four to three. For widescreen LCD displays, the aspect ratio is 16:9, very useful for viewing DVD movies, playing games and displaying multiple windows side by side. High-definition TV also uses this format. The viewable screen size is measured diagonally, so a 19" screen measures 19" from the top left to the bottom right.

Inside the computer there is a **video adapter**, or graphics card, which processes images and sends signals to the monitor. CRT monitors use a **VGA** (**video graphics adapter**) cable, which converts digital signals into analogue signals. LCD monitors use a **DVI** (**digital video interface**) connection.

Colour depth refers to the number of colours a monitor can display. This depends on the number of bits used to describe the colour of a single pixel. For example, an old VGA monitor with an 8-bit depth can generate 256 colours and a SuperVGA with a 24-bit depth can generate 16.7 million colours. Monitors with a 32-bit depth are used in digital video, animation and video games to get certain effects.

Display technologies

An **LCD** is made of two glass plates with a liquid crystal material between them. The crystals block the light in different quantities to create the image. **Active-matrix LCDs** use **TFT** (**thin film transistor**) technology, in which each pixel has its own switch. The amount of light the LCD monitor produces is called brightness or luminance, measured in cd/m^2 (candela per square metre).

A **CRT** monitor is similar to a traditional TV set. It contains millions of tiny red, green and blue phosphor dots that glow when struck by an electron beam that travels across the screen and create a visible image.

PCs can be connected to **video projectors**, which project the image onto a large screen. They are used for presentations and home theatre applications.

In a **plasma screen**, images are created by a plasma discharge which contains noble (non-harmful) gases. Plasma TVs allow for larger screens and wide viewing angles, making them ideal for movies.

Organic Light-Emitting Diodes (**OLEDs**) are thin-film LED displays that don't require a backlight to function. The material emits light when stimulated by an electrical current, which is known as electroluminescence. They consume less energy, produce brighter colours and are flexible – i.e. they can be bent and rolled up when they're not being used.

3 *Choosing the right display device*

A Listen to five customers in a computer shop describing their display device needs. Which device (a–e) would you recommend to each person? In pairs, discuss your choices and give reasons for them.

Speaker 1 .... Speaker 4 ....

Speaker 2 .... Speaker 5 ....

Speaker 3 ....

NEC MultiSyn LCD Monitor

Screen size: 17″

Resolution: 1280x1024 (SXGA)

Aspect ratio: 5:4

Brightness: 400 cd/m^2

a

Dell UltraSharp LCD monitor

Widescreen 24″ flat panel

Resolution: 1920x1200

Colour support: 16.7 million

Multiple video inputs, flash-card slots and USB ports

b

Cambridge-Hitachi interactive whiteboard

Allows interaction with a projected computer image

Board size: 78″

Connected to the PC via USB

Pointing device: cordless pen

c

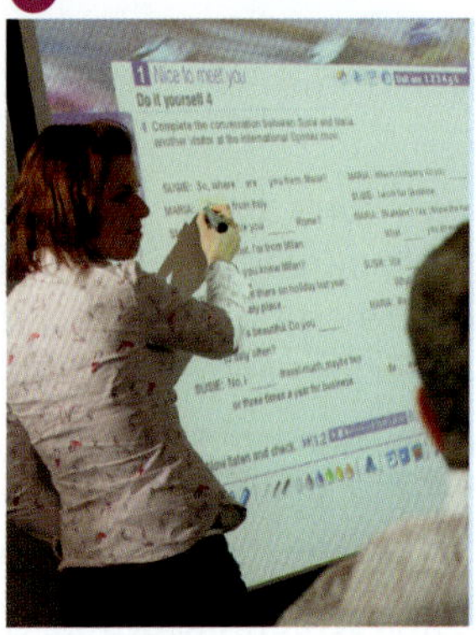

Pioneer 50″ Plasma TV

Resolution: 1280x768 (XGA)

Blu-ray Disc recorder

5.1 surround sound system (Five audio channels plus one subwoofer)

d

Portable projector

DLP (Digital Light Processing) technology

Resolution: 1024x768

Projection screen

e

B In pairs, discuss which of the display devices you would most like to own. Give reasons for your choice.

4 Ergonomics

A **Listen to Tony Clark, an expert in computer ergonomics, talking to some office workers about health and safety. What health problems associated with computer use do the office workers mention?**

B **Listen again and complete these extracts.**

1 Get a good chair, one that .... your lower back and is .... …

2 Make sure your feet rest firmly .... or on a footrest.

3 Position the keyboard .... your elbows, with your arms .... the work surface …

4 … position the monitor at, or just below, ...

5 You should sit at .... from the front of the monitor, about 50 to 70 centimetres away.

6 … a kind of stand that lets you move the monitor .... , so you can use it at the correct angle and height.

C **Match the extracts above (1–6) with the correct parts of the diagram (a–f).**

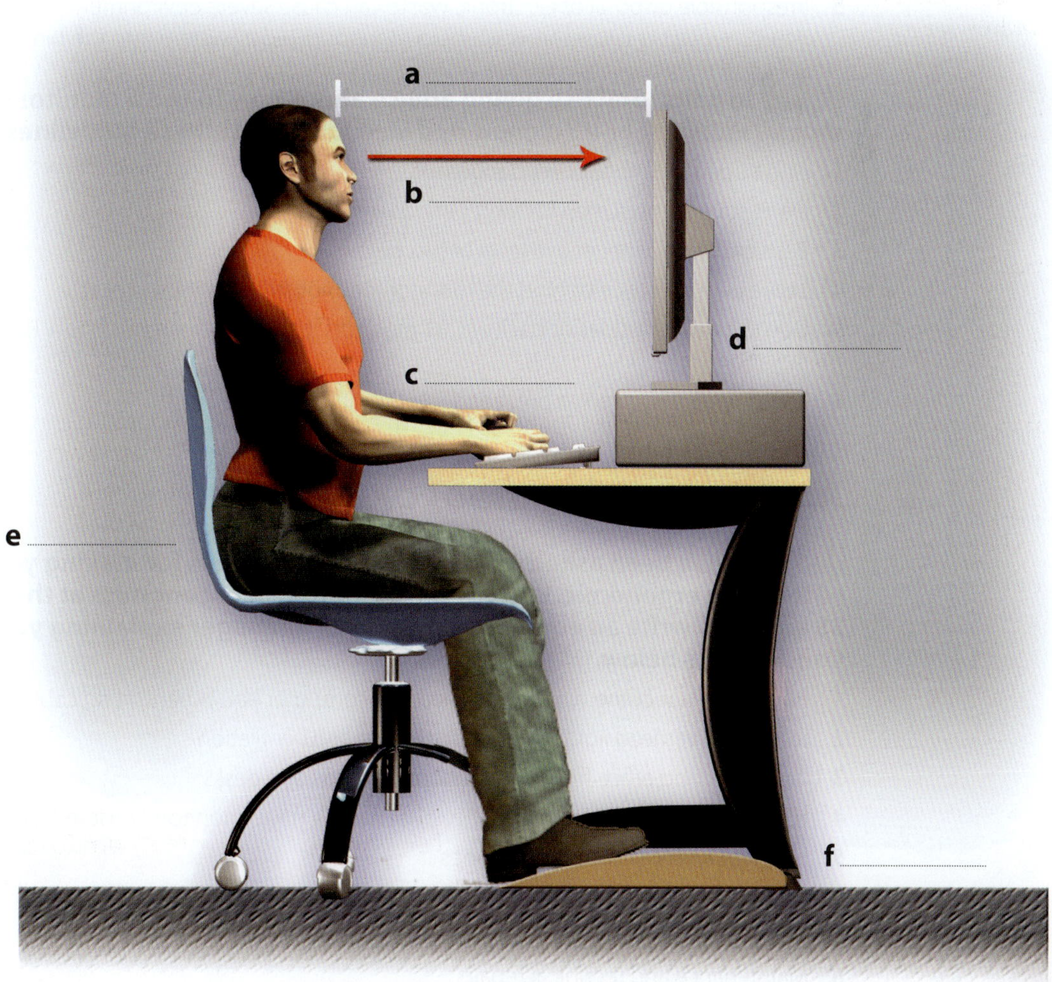

Ergonomics – the study of how people interact safely and efficiently with machines and their work conditions

5 Language work: instructions and advice

A Look at the HELP box and then complete these health and safety guidelines with *should/shouldn't*.

1 If you type a lot at your computer each day, you .... buy an ergonomic keyboard; it can help reduce the risk of repetitive strain injury.

2 You .... place your mouse within easy reach and support your forearm.

3 If you decide to build your own PC, protect yourself from electric shocks. You .... touch any components unnecessarily.

4 You .... always use a copyholder if you are working from documents. The best position is between the screen and the keyboard, or at the same height as the screen; this can reduce neck, back and eyestrain.

5 Irresponsible disposal of electronic waste can cause severe environmental and health problems. You .... just throw your old monitor or video system into the bin.

HELP box

Instructions and advice

- We use the imperative to give instructions.

  Get *an adjustable chair.*

  Don't put *your monitor in front of a window.*

- We use **should** and **shouldn't** + infinitive to give advice or to talk about what we think is a good or bad idea.

  You ***should*** *look down at the monitor, not up.*

  You ***shouldn't*** *use a monitor that's fuzzy or distorts the image.*

- We can also give advice by using set phrases like **It's a good idea to** or **It's a bad idea to** + infinitive.

  It's a good idea to *have a monitor with a tilt-and-swivel stand.*

B In pairs, practise giving advice about how to use a monitor safely using *should/shouldn't* or *It's a good/bad idea to*. Look at these guidelines for help.

1 Don't open the monitor. It's dangerous.

2 Don't stare at the screen for long periods of time.

3 Position the monitor at eye level or just below.

4 Leave enough space behind the monitor for unobstructed movement.

5 Don't sit near the sides or back of CRT monitors. Use LCD screens instead – they're free from radiation.

6 Keep the screen clean to prevent distorting shadows.

6 An ergonomic school or office

You have been asked to write a list of guidelines for making your school or office more ergonomic. Look at the definition of ergonomics at the bottom of page 35 and then write an email to your teacher/manager explaining your guidelines. Consider 1–8 below.

1 Physical layout of the work site: desk areas, computer equipment, filing cabinets, etc.

2 Lighting (overhead lights, desk lamps), glare and ventilation

3 Computer and office furniture: ergonomic chairs and desks

4 User-friendly and ergonomic devices: keyboards, mice, monitors, wrist rests, copyholders, etc.

5 Location and features of telephones

6 Layout of cables and switches for a wired network

7 Wireless internet access and wireless network

8 Maintenance and technical repairs

Choosing a printer

1 Types of printer

A How many types of printer can you think of? Make a list.

B Read the article on page 38 and then label the types of printer (1–5). Which types of printer aren't pictured?

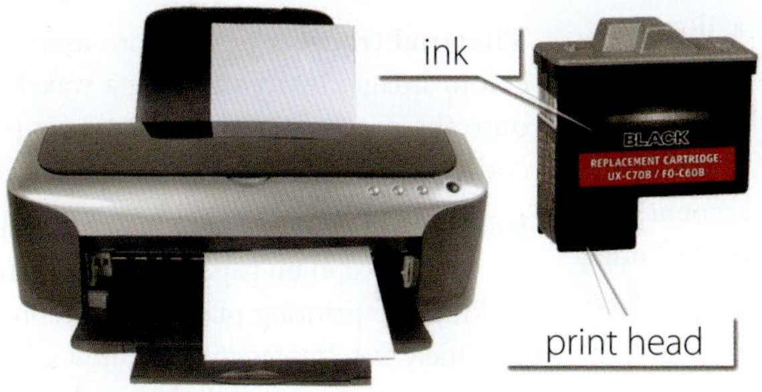

The quality (resolution) of the images goes up to 2,400 dots per inch (dpi)

1 ....

Provides high quality output: a resolution of 1,200–2,400 dpi

2 ....

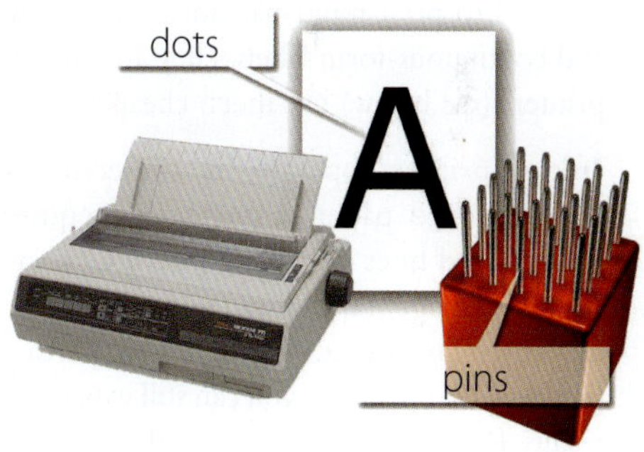

The resolution depends on the number of pins

3 ....

Provides high quality for linework (lines and curves)

4 ....

Provides the highest resolution: more than 3,000 dpi

5 ....

WHICH TYPE OF PRINTER SHOULD I BUY?

Printing is the final stage in creating a document. *Since* the results you can obtain with different types of printer will vary substantially, here is a guide to help you decide which one is most suitable for your needs.

To begin with, you should take into account that printers vary in cost, speed, print quality, and other factors *such as* noise or printing method. Technology is evolving so quickly that there is always a printer for every application or need.

Dot-matrix printers use pins to print the dots required to shape a character. They can print text and graphics; however, they produce relatively low resolution output – 72 to 180 dots per inch (dpi). They are used to print multi-part forms, self-copying paper and continuous-form labels. They are slower than laser printers (see below) but much cheaper.

Inkjet printers operate by projecting small ink droplets onto paper to form the required image. Colour and hues are created by the precise mixing of cyan, magenta, yellow and black inks. Inkjets are fairly fast, quiet, and not as expensive as laser printers. Nevertheless, you can still expect high quality results *because* there are some inkjet printers on the market with a resolution of 2,400 dpi.

Laser printers produce output at great speed and with a very high resolution of 1,200–2,400 dpi. They scan the image with a laser beam and transfer it to paper with a special ink powder called toner. They are constantly being improved. In terms of speed and image quality, laser printers are preferred by experts for various reasons; *for instance*, they have a wider range of scalable fonts than inkjets, can emulate different language systems, and can produce high-quality graphics; however, they are still expensive for home users.

Thermal transfer printers are used to produce colour images by transferring a wax-based ink onto the paper. They are popular for printing bar codes, labels and medium-resolution graphics.

Imagesetters produce very high-resolution output (up to 3,540 dpi) on paper or on the actual film for making the printing plates. In addition, they are extremely fast. Imagesetters are most often used in desktop publishing (DTP). Although they produce the highest quality output, they have one important disadvantage: they are too expensive for homes or small offices.

In modern lithographic printing, images are created on a DTP computer and *then* output directly to the printing plates, without requiring film as an intermediate step. This technology is called **computer to plate**, or **CTP**, and the machine used is called a **platesetter**.

Finally, we have **plotters**. Plotters use ink and fine pens held in a carriage to draw very detailed designs on paper. They are used for construction plans, engineering drawings and other technical illustrations. Nowadays, traditional plotters are being replaced with wide-format inkjets.

C Find words in the article with the following meanings.

1 designs and images used in magazines, books, etc. (lines 10–15) ....

2 output quality, measured in dots per inch (lines 10–15) ....

3 a particular colour within the colour spectrum (lines 15–20) ....

4 an ink powder used in laser printers and copiers (lines 25–30) ....

5 set of characters that can be resized (enlarged or reduced) without introducing distortion (lines 30–35) ....

6 a rectangular pattern of black lines of magnetic ink printed on an object so that its details can be read by a computer system (lines 35–40) ....

7 surface that carries a reproduction of the image, from which the pages are printed (lines 45–50) ....

8 in-between; middle (lines 50–55) ....

2 Language work: connectors 1

A Look at the HELP box and then put the words in italics from the article on page 38 into the correct column of the table.

Giving examples	Listing/Sequencing	Giving reason/cause

B Try to add some more connectors to each column. How do you say these connectors in your language?

HELP box

Connectors 1

Connectors are linking words and phrases which join ideas and help us organize our writing. Connectors can be used for giving examples, listing or sequencing, and giving reason or cause.

... ***for instance***, *they have a wider range of* ...

To begin with, *you should take into account that printers vary in cost* ...

Some common connectors appear in *italics* in the article on page 38. For more on other uses of connectors, see Unit 11.

C Write a paragraph describing the printer(s) you use at home or at work. Try to use some connectors. Think about these aspects: type, speed, resolution, print quality, memory, cost, print consumables (ink cartridges etc.).

3 Choosing the right printer

A In pairs, choose the most suitable printer for each of these situations. Give reasons for your choices.

1 You want to print documents, web pages and occasional photographs at home.
2 A small company needs a printer which will be shared by various users on a local area network (LAN).
3 A professional team of architects and engineers need to create accurate representations of objects in technical drawings and CAD.
4 A graphic arts business needs a printer to produce catalogues, brochures and other publications.

B In pairs, describe the features of your ideal printer.

4 Multi-function printers

A Listen to an extract from a consumer technology podcast about multi-function printers. What two disadvantages of multi-function printers are mentioned?

B Listen again and answer these questions.

1 What is a multi-function printer?
2 Why are multi-function printers so popular?
3 What is the main advantage of PictBridge technology?
4 Apart from sheets of paper, what other things can multi-function printers print?
5 What software do you usually get when you buy a multi-function printer?
6 What advice does Mr Kelly give on ink cartridges?
7 What type of device does he recommend for home users?
8 What type of device does he recommend for businesses?

5 Language work: comparatives

Look at the HELP box and then complete these sentences using the comparative form of the adjective in brackets.

1 A laser printer is generally (quiet) .... than a low-cost inkjet printer.
2 Multi-function printers are now only slightly (expensive) .... than conventional printers, and offer much (great) .... versatility.
3 The print quality of this network printer is noticeably (good) .... than any inkjet, and as (good) .... as similar laser printers.
4 The Agfa platesetter is (reliable) .... and (easy) .... to use than most printers of its type.
5 Your printer is only as (good) .... as the paper you use.
6 The final result is always (accurate) .... than the original image.
7 An imagesetter is (heavy) .... than a laser printer.

HELP box

Comparatives

- We form the comparative of one-syllable adjectives by adding **-er**.

  *slow → **slower***

  *Inkjet printers are **slower** than laser printers, but much **cheaper**.*

- Two-syllable adjectives usually take **more/less**.

  *modern → **more** modern*

  *They're designing a **more modern** version at the moment.*

- Adjectives ending in **-y** (for example, **noisy**) take **-er** and the **y** changes to **i**.

  *Dot-matrix printers are **noisier** than inkjets.*

- We form the comparative of adjectives with three or more syllables by adding **more/less**.

  *versatile → **more** versatile*

  *... they're cheaper and **more versatile** than standalone products.*

- Note the irregular forms:

  *good → **better***
  *bad → **worse***
  *little → **less***

  *If you want **better** results, you'll need specialized software.*

- Equality is expressed by using **as** ... **as**. Difference can be shown by using **not as** ... **as**.

  *This is **as fast as** many other printers in its class.*

  *Inkjets are **not as expensive as** laser printers.*

6 Reading quiz – printer adverts

A In pairs, read the adverts and then answer these questions. See who in your class can finish first.

1. How many inkjet printers are advertised?
2. Which printer would you recommend to someone who wants to print advertising graphics?
3. If you have the wide-format printer from Vutek, what kinds of material can you print on?
4. Which technology lets you print directly from your digital camera without needing a computer in between?
5. A page description language, or PDL, describes how to print the text and pictures on the page. Can you find two laser printer languages?
6. What is the resolution of the Brother HL Network Colour Laser Printer?
7. How fast is the Brother HL Network Colour Laser Printer?

Canon Compact Photo Printer SELPHY CP750 Photo Printer

An inkjet photo printer with a 2.4" colour LCD for easy viewing, editing and printing of perfect borderless photos. With PictBridge, you can print directly from digital cameras, memory cards or camera phones (via IrDA or optional Bluetooth unit) without connecting to a PC.

Resolution: 300x300 dpi

Software: Easy-PhotoPrint

Dimensions: 179x127.1x63 mm

Weight: 960g

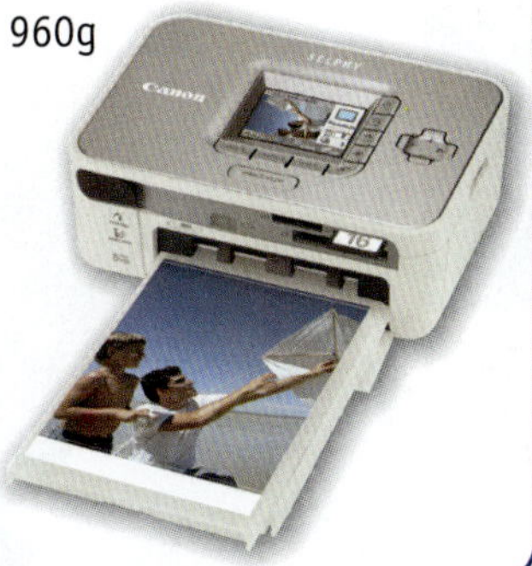

The Vutek UltraVu II 5330 provides the ultimate combination of highest print speed and best print quality in a five-metre printer.

- Wide-format professional inkjet printer
- Prints on a wide variety of substrates, including vinyl, and pressure-sensitive paper, mesh and textiles
- VUTEK Low Friction Kit allows for difficult materials to be run more easily
- Prints up to 16.4 feet (5 metres) wide
- Up to 330 dpi resolution produces images that are sharp, crisp and consistent
- Prints up to 2,230 square feet (207 square metres) per hour
- Applications: banners, exhibition graphics, bus shelters, etc.

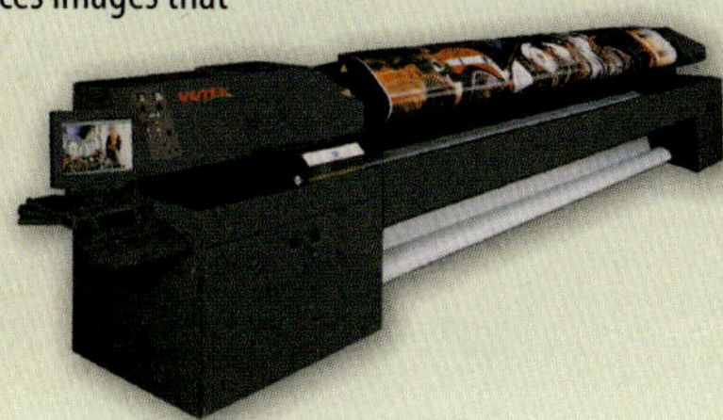

Brother HL Network Colour Laser Printer

The HL-4040CN delivers the perfect balance of quality, workgroup, colour A4 laser printing.

It boasts outstanding colour output: 2,400 dpi class colour printing with exceptionally crisp, high-resolution text and graphics driven by Brother's exclusive printing enhancement technologies.

Print Speed: up to 31 ppm (pages per minute) mono, 8 ppm colour (A4)

Compatibility: PCL and PostScript languages

Paper tray capacity: 250 sheets

Memory size: 64MB

High-speed USB

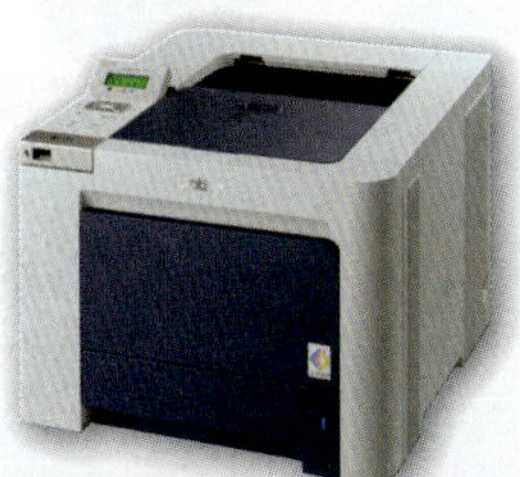

B A friend has emailed you asking for advice about which printer to buy, the Canon SELPHY CP750 or the Brother HL Network Colour Laser Printer. Write an email to your friend comparing the two printers. Use the HELP box on page 40 to help you.

Devices for the disabled

1 Assistive technology

A **In pairs, look at the words in the box and use as many of them as you can to describe the photos. You will not need all the words.**

blind person	adapted keyboard
motor-impaired person	on-screen keyboard
screen magnifier	voice recognition system
Braille printer	screen-pointing device
adaptive switch	screen reader
touch screen	pneumatic switch (sip and puff)

a

b

c

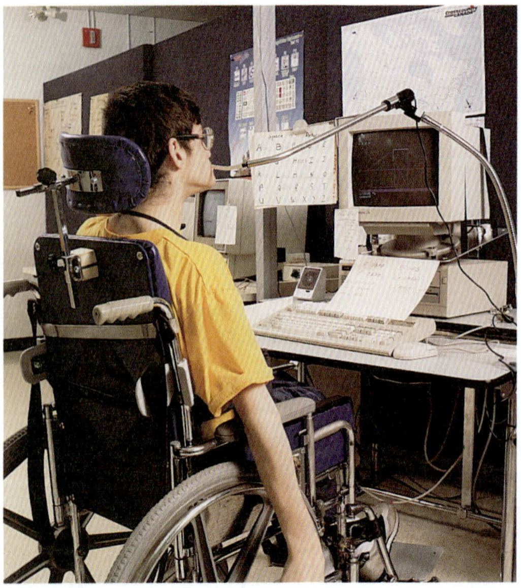

d

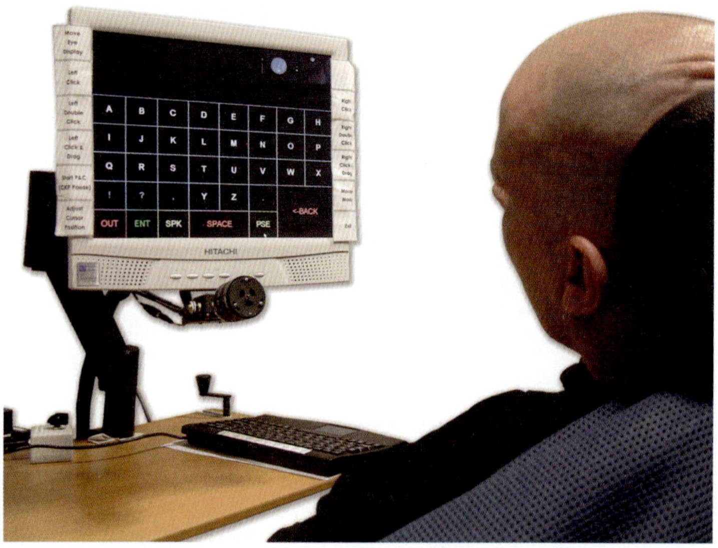

B **In pairs, discuss these questions.**

1 What sort of difficulties do you think are experienced by computer users with limitations of vision or mobility?

2 What types of device could be helpful to blind users?

3 How can a person with mobility limitations communicate with a computer?

2 Computers for the disabled

A **Read the text and find the following.**

1 the laws which ensure equal opportunities for people with disabilities in the USA and the UK

2 how the blind student in the photo interacts with the machine

3 the systems which type on the screen what is being said in meetings

4 the type of software which reads printed material, recognizes the text and then sends it to the PC

5 the system which is activated by the user's eye movements

6 the switch which can be used by someone with quadriplegia

7 the function of voice recognition devices

Computers for the disabled

Computers have taken a dominant role in our society, meaning most jobs now require access to computers and the Internet. But what happens if a person is blind, deaf or motor-disabled? They needn't worry. The latest assistive technology is designed to help them use computers and do their jobs in the office, learn at school, or interact with their families at home. In addition, new laws oblige companies to adapt the workplace to accommodate disabled people. For example, the Americans with Disabilities Act (ADA) and the UK's Disability Discrimination Act make it illegal for employers to discriminate against people with disabilities.

To work effectively, most blind users need to have their computers adapted with technologies such as **Braille, screen magnifiers, speech synthesis** and **Optical Character Recognition** (OCR).

Braille keyboards have Braille lettering on keyboard overlays, allowing the blind user to easily identify each key. For output, there are printers, called **Braille embossers**, that produce tactile Braille symbols on both sides of a page at high speed.

For someone with limited but usable vision, a screen magnifier may be appropriate. This type of software can enlarge text and images appearing on the screen by up to 16 times.

A Braille embosser prints a hard copy of a text document in Braille

A speech synthesis system is used to read aloud the work on the computer. It has a speech synthesizer, which produces the audio output, and a screen reader – the program which reads aloud text and menus from word processors, databases and the Web.

OCR uses a flatbed scanner and specialized OCR software to read printed material and send the text to the computer. The PC can then produce a copy of the text in Braille, a magnified copy, or a version that can be read aloud by a speech synthesis system.

Deaf computer users can overcome many communication difficulties with the aid of **visual alerts**, **electronic notetakers** and **textphones**. Visual alerts are indicators that alert the deaf user when they receive new mail or when there is a system error. So instead of hearing a sound, the user is alerted by a blinking menu bar or by a message on the screen. Electronic notetakers use software that types a summary of what is said in meetings onto the computer screen.

Textphones allow the deaf to type and read phone conversations. They are also called **TDDs** (Telephone Devices for the Deaf) or **TTYs** (TeleTypewriters). They can be used in combination with relay services, where an operator says what the text user types, and types what a voice phone user says. Deaf people can also communicate via SMS and instant messaging.

A textphone

Motor-impaired workers unable to type on a standard keyboard can employ **expanded** or **ergonomic keyboards**, **on-screen keyboards**, **adaptive switches** and **voice recognition systems**.

A specialized keyboard for children with physical disabilities

On-screen keyboards are software images of a keyboard that appear on the screen and may be activated with a trackball, touch screen, screen-pointing device, or eye movements. In an **eyegaze system**, the keys on the virtual keyboard are activated by the user's eyes when they pause on a key for two or three seconds.

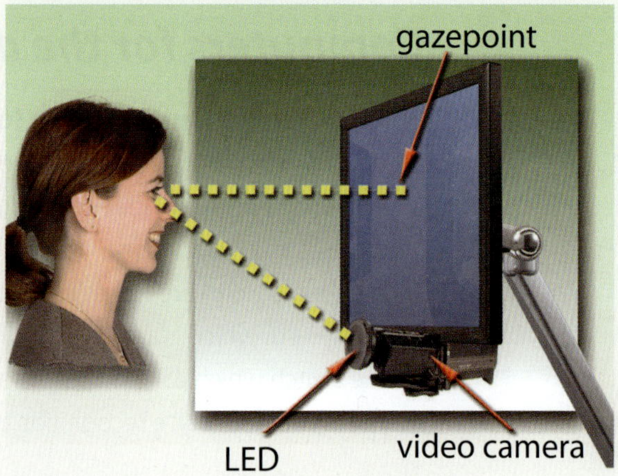

Eyegaze technology consists of a video camera and image processing software, which determines the eye's gazepoint on the screen

Switches come in many shapes and sizes. They are operated by muscle movements or breath control. For example, a **pneumatic switch** – known as a **sip and puff** – allows someone with quadriplegia to control the PC by puffing and sipping air through a pneumatic tube. People with quadriplegia can also use sip and puff joysticks.

Finally, there's voice recognition, which allows the computer to interpret human speech, transforming the words into digitized text or instructions.

B Complete the crossword with words from the text on pages 43–44.

ACROSS

2 An .... keyboard presents a graphic representation of a keyboard on the desktop screen and allows people with mobility impairments to type data using a joystick or a pointing device.

4 Visual .... allow deaf users to be notified of incoming mail or error messages without hearing a tone.

6 A screen .... makes the computer screen more readable for users with poor vision.

7 A system of reading and writing using raised dots, which enables blind people to read by touch.

DOWN

1 Unlike a standard telephone, a .... has a small screen and a keyboard that transcribes a spoken voice as text. It is used for text communication via a telephone line, ideal for people who have hearing or speech difficulties.

3 A Braille .... is an impact printer that prints text as Braille, by punching dots onto paper.

5 A speech synthesizer is used in conjunction with a screen .... to convert screen contents into spoken words.

3 *Language work: noun phrases*

A Look at the HELP box and then the noun phrases 1–6. Decide what type of modifier (a–d) is placed before the 'head' in each case.

1 disabled worker
2 rehabilitation engineer
3 employee's abilities
4 adapted keyboard
5 voice-activated computer
6 pointing device

a adjective
b present participle
c 's genitive
d noun

B Explain the noun phrases in A.

Example:
disabled worker = *a worker who is disabled*

HELP box

Noun phrases

A noun phrase is a phrase that has a noun as its head. This noun can be accompanied by a modifier that gives information about the head.

modifier	**head**
speech	*recognition*
compatible	*computer*

A noun phrase can function as the subject or object of a verb. It can contain the following range of modifiers:

- adjectives
  I have a portable computer.
  = a computer which is portable
- present participles
  I use this drawing program.
  = a program that draws
- **'s** genitive
  The files are on the director's computer.
  = the computer which belongs to the director
- nouns
  I need to buy a colour scanner.
  = a scanner which uses colour

4 *Assistive technologies for the blind*

A Listen to an interview with Mike Hartley, the director of the Assistive Technology Project for the Blind in Washington DC. Make notes about these topics.

1 The work that Mike is currently involved in:

2 Assistive technologies for blind users:

3 The difference between voice recognition and speech synthesis:

4 The goal of the Web Accessibility Initiative:

5 Companies developing assistive technology products:

A Braille computer keyboard

Small-Talk Ultra, a talking computer from GW Micro, which includes Window-Eyes – a screen reader for the blind

B In pairs, help each other to improve your notes and then listen again to make sure you have included all of the important information.

5 *Investing in assistive technologies*

Your school/company has decided to invest some of its annual IT budget in assistive technology. Write an email to your director of studies / manager, summarizing the different technologies available and the kind of people they can help. If possible, use the Internet to find suppliers of these technologies in your country.

Now visit www.cambridge.org/elt/ict for an online task.

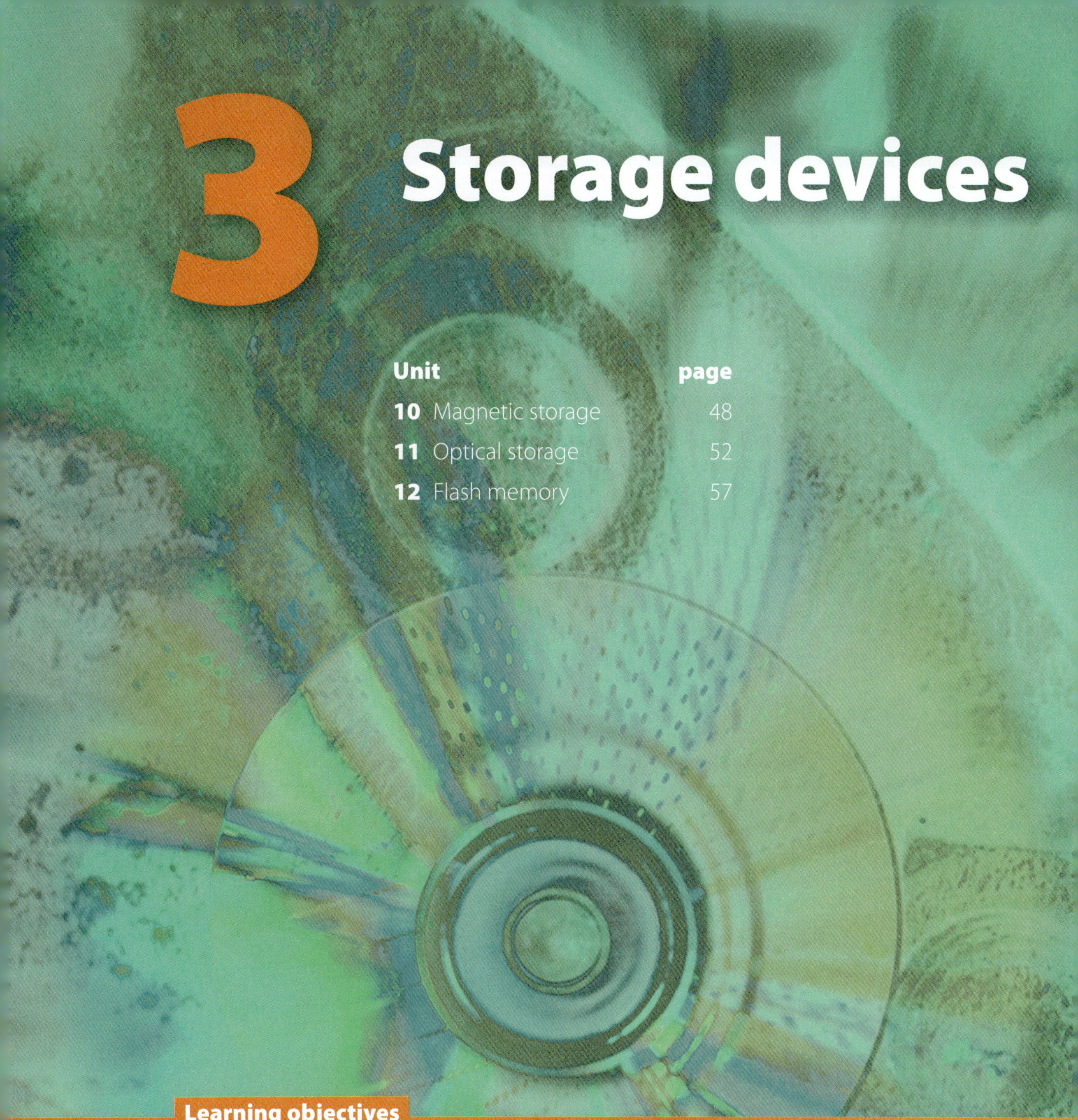

3 Storage devices

Learning objectives

In this module, you will:

- learn about different types of magnetic drive and disk.
- give instructions and advice on how to protect data.
- use technical vocabulary associated with optical storage devices and media.
- learn and use more discourse connectors.
- learn about the technical details of flash memory and its uses.
- learn different ways of making new words: affixation, conversion and compounding.
- describe flash-based devices.

Unit 10

Magnetic storage

1 *Types of magnetic drive*

A Look at the pictures and descriptions below and find the following.

1 the name of the hard drive on a PC platform
2 the type of hard drive that plugs into a socket at the back of a computer
3 the system that works in sequential format
4 the size and storage capacity of a floppy disk

A 3.5" floppy drive and diskette

A floppy disk drive uses 3.5" disks, which can store 1.44MB of data; it is usually assigned to the A: drive. Floppy drives are becoming increasingly rare.

The inside of a hard drive

Most PCs have one internal hard drive, usually called C: drive. It is used to store the operating system, the programs and the user's files in a convenient way. A hard drive can hold hundreds of gigabytes of data.

A portable external hard drive

External hard drives are connected to the USB or FireWire port of the computer. They can be as small as a wallet but can have as much capacity as internal drives; they are typically used for backup or as secondary storage.

Magnetic tapes and drive

A tape drive reads and writes data on tapes. It is sequential-access – i.e. to get to a particular point on the tape, it must go through all the preceding points. Tapes can hold hundreds of gigabytes of data and are used for data collection, backup and archiving.

B Complete these sentences with words from the box.

capacity	storage	archiving	hold	secondary

1 There are basically three types of magnetic .... device available to the computer user – hard drives, diskettes and tapes.
2 The .... of a 3.5" floppy disk is only 1.44MB.
3 Hard drives can .... hundreds of times more data than floppy disks.
4 A portable hard drive is a good choice for .... storage.
5 Magnetic tapes are used for .... information that you no longer need to use regularly.

2 *Buying a portable hard drive*

A Sue (see Unit 4) wants to buy a new drive. Listen to her conversation with the sales assistant. Does she buy anything?

B Listen again and answer these questions.

1 What is the storage capacity of the Iomega eGo portable hard drive?
2 How much information can be stored on the Edge DiskGo model?
3 Which hard drive is good for mobile professionals?
4 How much does the Iomega eGo drive cost?
5 How much does the Edge DiskGo cost?

The Iomega eGo portable hard drive.

3 Magnetic storage

A Read the text and then identify a sector and a track in Fig. 1.

B Read the text again and decide whether these sentences are true or false. Correct the false ones.

1 A hard drive spins at the same speed as a floppy disk drive.
2 If you format a hard drive that has files on it, the files will be deleted.
3 Hard drives cannot be partitioned to run separate operating systems on the same disk.
4 *Seek time* and *transfer rate* mean the same thing.
5 Disk drives are not shock resistant, especially in operating mode.

Magnetic storage

Magnetic storage devices store data by magnetizing **particles** on a disk or tape.

A **floppy disk** is so called because it consists of a flexible sheet of plastic, coated with iron oxide – a magnetizable material. A floppy disk drive spins at 360 revolutions per minute (rpm), so it's relatively slow. However, a **hard drive** spins at over 7,200 rpm and stores data on a stack of metal rotating disks called **platters**. This means you can store much more data and retrieve information much faster.

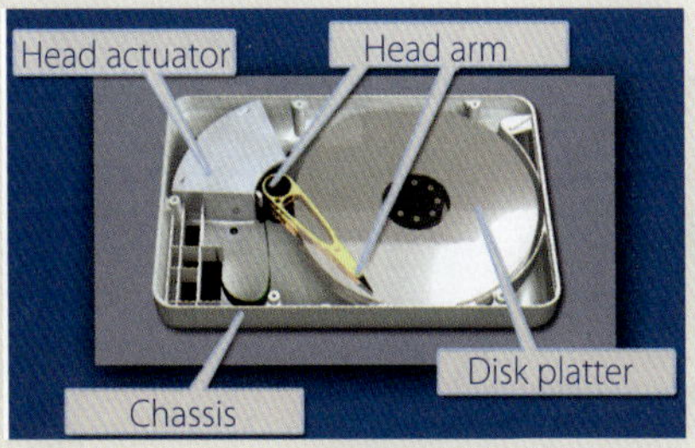

The inside of a hard drive

New disks need to be **formatted** before you can use them, unless they come preformatted from the manufacturer. When the disk is formatted, the operating system (OS) organizes the disk surface into circular **tracks** and divides each track into **sectors**. The OS creates a **directory** which will record the specific location of files. When you save a file, the OS moves the **read/write head** of the drive towards empty sectors, records the data and writes an entry for the directory. Later on, when you open that file, the OS looks for its entry in the directory, moves the read/write heads to the correct sector, and reads the file in the RAM area. However, formatting erases any existing files on a disk, so do not format disks on which data that you don't want to lose is stored.

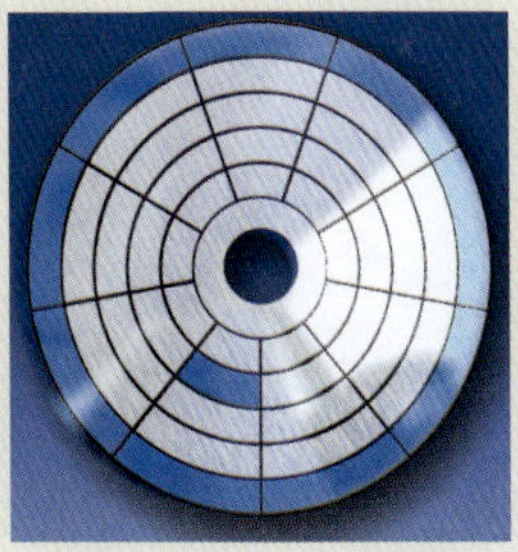
Fig. 1

The OS allows you to create one or more **partitions** on your hard drive, in effect dividing it into several logical parts. Partitions let you install more than one operating system (e.g. Windows and Linux) on your computer. You may also decide to split your hard drive because you want to store the OS and programs on one partition and your data files on another; this allows you to reinstall the OS when a problem occurs, without affecting the data partition.

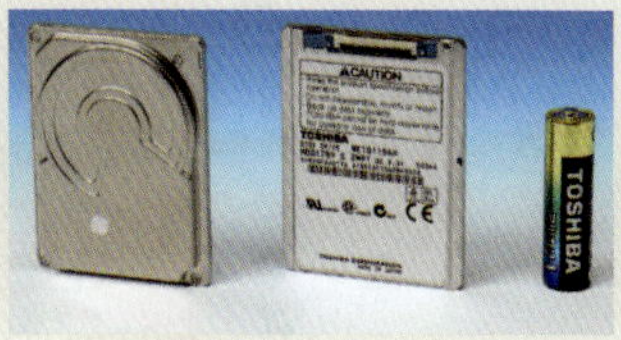

Toshiba's 1.8" hard drive; mini hard drives are used in small gadgets, such as PDAs and wristwatches

The average time required for the read/write heads to move and find data is called **seek time** (or **access time**) and it is measured in milliseconds (ms); most hard drives have a seek time of 7 to 14 ms. Don't confuse this with **transfer rate** – the average speed required to transmit data from the disk to the CPU, measured in megabytes per second.

How to protect your hard drive

- Don't hit or move the computer while the hard drive is spinning. Hard drives are very sensitive to vibration and shocks, especially when they are operating; when the read/write head touches the rotating disk, it can scratch and damage the disk surface. This is known as **head crash**.
- You shouldn't turn your computer off and on quickly. Wait at least ten seconds to ensure that the drive has stopped spinning.
- Check your hard drive regularly for logical and physical errors. To check and repair a drive, you can use a disk diagnosis utility like Windows ScanDisk.
- To minimize the risk of data loss or corruption, you should install an up-to-date virus scanner. You should also **back up** your hard drive regularly.

C Match these words (1–5) with the definitions (a–e).

1	formatted	a	a file system that defines the structure for keeping track of the files
2	directory	b	the part of a drive that reads and records data on a disk
3	read/write head	c	to make a copy of data or software in case the original disk is damaged
4	head crash	d	initialized; when the tracks and sectors on magnetic disks are set
5	back up	e	a serious disk malfunction; when the read/write head touches the rotating disk

4 *Language work: precautions*

A Look at the HELP box and then match the instructions (1–6) with the pictures (a–f).

1. Do not expose discs to heat or direct sunlight.
2. Check for viruses before opening files you receive from the Web or via email.
3. Make backup copies of your files.
4. Don't shake or move the computer violently while the hard drive is spinning.
5. Keep your discs away from water and humidity.
6. Hold discs by the edges, or by one edge and the centre hole.

a

d

b

e

c

f
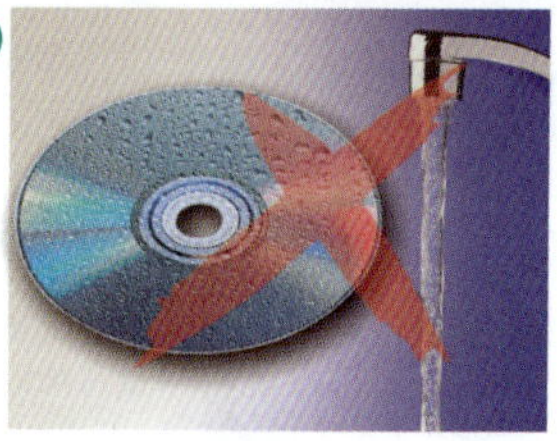

HELP box

Precautions

- We use the imperative to give precautions and warnings.

  ***Check** your hard drive regularly for logical and physical errors.*

  *... formatting erases any existing files on a disk, so **do not format** disks on which data that you don't want to lose is stored.*

- We use **should** + infinitive without *to* to give advice or to talk about what we think is right.

  *... you **should** install an up-to-date virus scanner.*

- We use **shouldn't** + infinitive without *to* to give advice or to talk about what we think is wrong.

  *You **shouldn't** turn your computer off and on quickly.*

B **In pairs, discuss what you should or shouldn't do to protect your data. Use the suggestions below.**

Example: discs on top of each other (stack)
You shouldn't stack discs on top of each other. / Don't stack discs on top of each other.

1 your anti-virus program regularly, since new viruses are created everyday (update)
2 discs in a protective case (store)
3 passwords and security devices to protect confidential information (use)
4 on discs with permanent marker pens (write)
5 the disc into the disc drive carefully (insert)
6 floppies or hard drives near magnets; they can damage the data stored on them (leave)

Note: dis**c** (optical media); dis**k** (magnetic storage media)

5 Word building

Look at the words in the boxes. Are they nouns, verbs, adjectives or adverbs? Write *n*, *v*, *adj* or *adv* next to each word and then complete the sentences below. For more about word building, see Unit 12.

magnet	magnetic	magnetically ....
magnetism	magnetize	magnetized ....

1 .... is the science of magnetic phenomena and properties.
2 Floppy disks and hard drives are .... storage devices.
3 Data is recorded on a disk in the form of .... spots called *bits*.

fragment	fragmentation ....
defragmenter	fragmented ....

In a fragmented disk, a file is stored in non-continuous sectors

4 After you create, delete and modify a lot of files, the hard drive becomes ...., with bits and pieces spread all over the disk.
5 .... slows down the speed at which data is accessed because the disk drive has to work harder to find the parts of a file stored in many different locations.
6 To reorganize your hard drive, you can use a disk optimizer or ....; this will reorder your files into continuous clusters.

In a defragmented disk, a file is stored in neighbouring sectors

6 Explaining hard drive precautions

A friend has sent you an email explaining that she has just lost all of the information on her PC because of a head crash. Write a reply explaining the following.

- Why the head crash happened
- What precautions she should take with her new PC to avoid similar problems in the future
- What steps she could take to back up her files

Unit 11 Optical storage

1 CDs and DVDs

A In pairs, discuss these questions.

1 What do CD and DVD stand for?

2 What is the main advantage of using DVDs instead of CDs?

B How do you say these expressions in your language?

1 optical disc

2 laser beam

3 backward-compatible

C Paul (see Unit 4) wants to buy some blank discs. Listen to his conversation with the sales assistant and check your answers to A.

D Listen again and decide whether these sentences are true or false. Correct the false ones.

1 A DVD is an optical digital disc that can be used for video, audio and data storage.

2 The dimensions of a CD and a DVD are the same: 1.3 mm thick and 13 cm in diameter.

3 The data on a DVD is read with a laser beam.

4 A basic DVD can hold 3.7 gigabytes.

5 You need a hard drive to read DVDs.

6 DVD-Video discs can hold full-length movies.

7 A DVD Writer is not compatible with old CD-ROMs.

A DVD drive with disc

Note: dis**c** (optical media); dis**k** (magnetic storage media)

2 Optical discs and drives

A Read the text on page 53 and find the following.

1 the advantages and disadvantages of optical discs over magnetic disks

2 the storage capacity of a double-sided, dual layer DVD

3 the difference between a DVD burner and a DVD recorder

4 the feature of a portable DVD player which allows the user to play different formats

5 two possible successors to DVDs

6 where the Blu-ray format gets its name from

Optical discs and drives

Optical discs can store data at much higher densities than magnetic disks. They are therefore ideal for multimedia applications where images, animation and sound occupy a lot of disc space. Furthermore, optical discs are not affected by magnetic fields, meaning that they are secure and stable, and can be transported through airport metal detectors without damaging the data. However, optical drives are slower than hard drives.

CDs and DVDs

At first sight, a **DVD** is similar to a **CD**. Both discs are 120 mm in diameter and 1.2 mm thick. They also both use a **laser beam** to read data. However, they are very different in internal structure and data capacity. In a DVD, the **tracks** are very close together, thus allowing more tracks. The **pits** in which data is stored are also smaller, so there are more pits per track. As a result, a CD can hold 650-700MB, whereas a basic DVD can hold 4.7GB. In addition, a DVD can be **double-sided** and **dual layer**, with a capacity of 17GB.

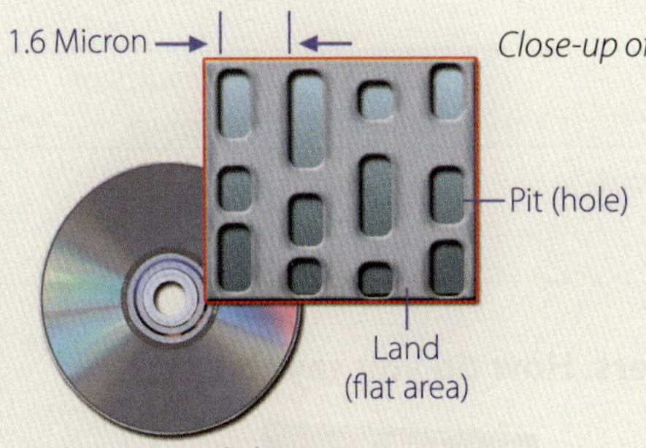

Close-up of a CD

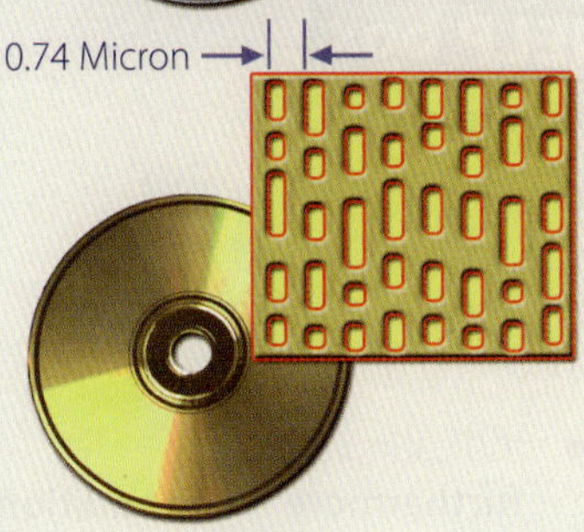

Close-up of a DVD

CDs come in three different formats:

- CD-ROMs (**r**ead-**o**nly **m**emory) are read-only units, meaning you cannot change the data stored on them (for example, a dictionary or a game).
- CD-R (**r**ecordable) discs are write-once devices which let you duplicate music CDs and other data CDs.
- CD-RW (**rew**ritable) discs enable you to write onto them many times, just like a hard disk.

DVDs also come in several formats:

- DVD-ROMs are used in DVD computer drives. They allow for data archiving as well as interactive content (for example, an encyclopedia or a movie).
- DVD-R or DVD+R can only be recorded on once.
- DVD-RW or DVD+RW discs can be erased and re-used many times. They are used to back up data files and to record audio and video.

The DVD drive used in computers is also called a **DVD burner** because it records information by burning via a laser to a blank DVD disc. However, a **DVD recorder** typically refers to a standalone unit which resembles a video cassette recorder. New DVD recorders can play all CD and DVD formats. There are also **portable DVD players** – handheld devices which let you watch movies or TV, play games and listen to music, wherever you are. They come with a built-in DVD drive and widescreen (rectangular 16:9 format) LCD display. They usually support **multi-format playback** – that is, they can play many file formats, including DVD-video, DivX, CD audio discs, MP3 music and JPEG images.

HD-DVD and Blu-ray discs

These two competing formats are expected to replace current DVD as the standard for watching movies at home. On one side are Toshiba, Microsoft and the DVD Forum, who support the **High Definition-DVD** (**HD-DVD**). Sony, Panasonic, Samsung, JVC and many movie studios are behind the **Blu-ray** format.

A Sony Blu-ray disc

A Blu-ray disc has a capacity of 25GB (single layer), 50GB (dual layer) and 100GB (four layer). Unlike DVDs, which use a red laser to read and write data, Blu-ray uses a blue-violet laser, hence its name. Blu-ray discs can record and play back high-definition television and digital audio, as well as computer data.

B **Read the text again and make notes about the features of CDs, DVDs and Blu-ray discs.**

	Capacity and formats	Possible uses
CD		
DVD		
Blu-ray		

3 *Language work: connectors 2*

A **Look at these extracts from the text and put the words in *italics* into the correct column of the table.**

1 They are *therefore* ideal for multimedia applications …
2 **Furthermore**, optical discs are not affected by magnetic fields.
3 **However**, they are very different in internal structure and data capacity.
4 **As a result**, a CD can hold 650–700MB, **whereas** a basic DVD can hold 4.7GB.
5 **In addition**, a DVD can be double-sided and dual layer …

Indicating addition	Making contrasts	Explaining the results or effects of something

B **Look at the HELP box and check your answers. How do you say these connectors in your language?**

C **Choose the correct word in brackets to complete these sentences.**

1 (Although/Consequently) CDs and DVDs are similar in size and shape, their data structure is very different.
2 DVDs hold more data than CDs. The pits burnt into the disc are smaller than on a CD, and the tracks are closer together. (On the other hand / As a result), DVDs can have up to four recording layers.
3 A Blu-ray disc drive costs a lot of money (but/so) you should use it carefully.
4 Blu-ray is expected to replace DVD over the coming years (because/besides) it offers much greater storage capacity.
5 Both Blu-ray (and / in addition) HD-DVD devices are backward-compatible with current CDs and DVDs, meaning you can play your old discs on the new players.
6 Sony has invested millions of dollars in the development of Blu-ray technology. The success of Blu-ray is (whereas/ therefore) vital for the company's future.

HELP box

Connectors 2

In addition to the uses of connectors covered in Unit 8, we also use connectors for the following purposes:

- Indicating addition
  furthermore **in addition**
  besides **moreover**
  and
- Making contrasts
  however **whereas**
  although **but**
  on the other hand
- Explaining the results or effects of something
  therefore **as a result**
  so **thus**
  consequently **because**

4 Choosing storage devices

In pairs, look at the products in the computer catalogue and choose the most suitable device for the purposes (1–6). Give reasons for your choices. Try to use some connectors from the HELP box on page 54.

1 to keep the operating system and the programs on a home computer
2 to watch a movie on a plane or in the back seat of a car
3 to hold your favourite photos and music
4 to make backup copies and to transport files between computers in a big company
5 to hold historical records in the National Library
6 to read, write and re-write high-definition video and TV

Seagate hard drive
Superfast 8ms hard drive. Capacity ranges from 80GB to 1TB.

Iomega portable hard drive
160GB, 2.5" external hard drive. An affordable way to back up all your data, from business documents to emails.

LaCie DVD drive
16x DVD writer with free Nero DVD burning software. Can play and record both DVD+R and DVD-R discs, plus their rewritable counterparts, as well as all types of CD.

Panasonic portable DVD player
8" portable LCD DVD Player with Car Kit. Compatible with DVD-Video, CD, JPEG image CD and MP3-formatted audio CD.

Sony Blu-ray disc drive
Sony's Vaio AR laptop is the first portable Blu-ray studio, which includes a Blu-ray disc drive and a TV tuner, alongside a 17" widescreen display and a 2GHz Intel Core Duo processor.

Toshiba USB flash drive
High-speed 16GB pen drive with a built-in MP3 player. Plugs directly into any USB connection.

Sony's Vaio AR laptop

Useful language

For this use, the … is the most appropriate because …

The … has … so I'd choose it for …

However, … is good for … because …

In a big company, it would be a good idea to …

Well, that depends on …

I agree / I disagree.

5 Format wars

Read these posts from a forum about the topic of 'Blu-ray versus HD-DVD' and then add your response, giving your opinion on the topic.

HD-DVD and Blu-ray formats display movies in full high-definition resolution, but they are incompatible; HD-DVD cannot play the Blu-ray discs, and vice versa. People say that Blu-ray discs can hold more data and video, but that they are more expensive and complex. Who will be the winner in this format battle?

Consumers ezine, March 10th at 5:40 pm

Samsung and Toshiba are selling hybrid players that can play both formats. Sony and NEC are also releasing dual-format players. This may be the end of the format war. Will both sides produce a unified standard?

News.net, March 15th at 12:30 am

I hate format wars. This situation reminds me of the Beta versus VHS war in the early days of the video market, and more recently DVD-R versus DVD+R. I don't want to invest money in equipment that quickly becomes obsolete or incompatible. Why can't someone create a universal player that plays all formats, from CDs to high-definition video discs?

Posted by Adam, March 15th at 4:15 pm

Name:

Submit comment

Unit 12 Flash memory

1 *Flash-based gadgets*

Flash memory is used in many handheld devices. Match the descriptions (1–6) with the pictures (a–f).

1 This handheld console lets you play games stored on ROM game cards, which have a small amount of flash memory to save user data, for example high scores.

2 This flash memory card is used as 'digital film' to store images on a digital camera.

3 This wireless LAN card allows laptop and PDA users to access the Internet from any Wi-Fi access point.

4 This USB flash pen drive is the latest mobile drive for your computer.

5 It looks like an ordinary watch, but this USB drive from Edge Tech can store up to 1GB of flash memory. It will let you save and transfer your photos, songs and data files easily.

6 This flash-based player provides everything you need to play music and store data on the go. It also comes with a built-in FM radio and voice recorder.

a ....

b ....

c ....

d ....

e ....

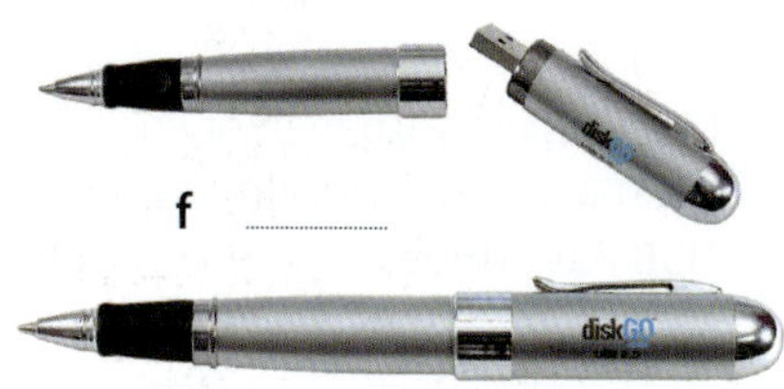

f ....

2 *Memory in a flash!*

A Look at the title of the text on page 58. Why is it a suitable title for an article about flash memory? Read the first paragraph of the text to find out.

B Read the whole text and answer these questions.

1 What is flash memory?

2 What are the differences between RAM memory and flash memory?

3 What can devices which use multi-level cell technology do?

4 What are the differences between flash drives and external hard drives?

5 What is the advantage of using U3 technology in flash drives?

6 How much data can a flash memory card hold?

7 What is the name of the flash card created by Sony for its digital cameras?

Memory in a flash!

Flash memory is a type of **non-volatile** memory that can be electronically erased and reprogrammed. Its name was invented by Toshiba to express how much faster it could be erased – 'in a flash', which means 'very quickly'.

Unlike RAM, which is **volatile**, flash memory retains the information stored in the chip when the power is turned off. This makes it ideal for use in digital cameras, laptops, network switches, video game cards, mobile phones and portable multimedia players. In addition, it offers fast read access times (although not as fast as RAM), with transfer rates of 12MB per second. Unlike ROM chips, flash memory chips are rewritable, so you can update programs via software.

Inside the chip, data is stored in several floating gate transistors, called **cells**. Each cell traditionally stores one bit of data (1 = erased and 0 = programmed). New devices have a multi-level cell structure so they can store more that one bit per cell. The chips are constructed with either **NOR** or **NAND** gates. NOR chips function like a computer's main memory, while NAND works like a hard drive. For example, in a camera, NOR flash contains the camera's internal software, while NAND flash is used to store the images.

Flash memory is used in several ways:

- Many PCs have their BIOS (basic input/output system) stored on a flash memory chip so it can be updated if necessary.
- Modems use flash memory because it allows the manufacturer to support new protocols.
- **USB flash drives** are used to save and move MP3s and other data files between computers. They are more easily transported than external hard drives because they use **solid-state** technology, meaning that they don't have fragile moving parts that can break if dropped. However, USB flash drives have less storage capacity than hard drives.
- New **U3 smart drives** allow users to store both applications and data. They have two drive partitions and can carry applications that run on the host computer without requiring installation.
- **Flash memory cards** are used to store images on cameras, to back up data on PDAs, to transfer games in video consoles, to record voice and music on MP3 players or to store movies on MP4 players. They are as small as a stamp, and capacity can range from 8MB to several gigabytes. The only limitation is that flash cards are often not interchangeable between devices. Some formats include: CompactFlash, Secure Digital, MultiMedia Card, miniSD card, and xD-Picture Card. Sony has its own product called the Memory Stick, used in its digital still cameras, video camcorders and the PlayStation Portable. The photos stored in a digital camera can be offloaded to a computer via cable or wirelessly. Another option is to have a **flash card reader** permanently connected to your PC; you simply eject the card from the camera and put it into the reader instead of having to plug the camera in.

The future of hard drives may be **hybrid** hard drives. Hybrid hard drives combine a magnetic hard disk *and* flash memory into one device. This allows computers to boot, or start, more quickly, and also reduces power consumption.

SanDisk's card readers read and write to just about every flash memory card

C Find words or phrases in the text with the following meanings.

1 permanent; able to hold data without power (lines 1–5) ....

2 able to be rewritten many times (lines 10–15) ....

3 different sections of a disk drive or storage area (lines 40–45) ....

4 to make a copy of a file so that the original is not lost (lines 45–50) ....

5 transferred to another device (lines 60–65) ....

6 a peripheral device that reads and writes flash memory cards (lines 60–65) ....

7 a product that integrates two different technologies (lines 65–70) ....

3 Language work: word building

A Look at the HELP box and then, using affixation, conversion and compounding, try to make as many words as you can from *blog*, *mail* and *print*. Use a dictionary and the Internet to help you.

blog	mail	print
blogger (a person who writes a blog)	*to mail* (the verb form)	*printout* (the pages produced by the printer)

B Choose the correct word in brackets to complete this description of a digital voice recorder. Use a dictionary to help you.

Olympus WS-320M digital voice recorder

Slim, attractive, and highly functional, the Olympus WS-320M digital voice recorder packs 1GB of internal flash memory into its **1** (lighted/lightweight/lighten) housing, letting you record up to 277 hours of high-quality audio in WMA format. It's ideal for **2** (record/recordable/recording) notes or long lectures, interviewing people, or capturing song ideas before they disappear. As an added bonus, the WS-320M can store up to 266 WMA or MP3 songs for high-quality stereo **3** (player/playback/playoff).

The WS-320M features five separate file **4** (folds/folding/folders), capable of holding 199 files each, so you can organize nearly 1,000 files by subject, theme or other category. Users also have the choice of four recording modes: HQ for high-quality audio, LP and SP for extended recording times, and ST HQ for stereo recording. And thanks to the voice **5** (activation/activate/active) option, users don't need to press a single button to start recording – the WS-320M will record as soon as the built-in microphone picks up sound.

Perhaps the most convenient feature, however, is the built-in USB **6** (connector/connect/connected), which eliminates the need for a USB cable. Once this is connected, you can **7** (downloadable/download/upload) music files, images or documents from your PC, in effect turning the recorder into a small hard drive. You can even transfer voice recordings to your computer for **8** (store/storage/storeroom) or multimedia use.

HELP box

Word building

We can create new words from existing words in three main ways:

- Affixation (adding a prefix or suffix)

  Adding a prefix:
  volatile → ***non**-volatile*
  date → ***up**date*

  Adding a suffix:
  erase → *eras**able***
  install → *install**ation***

- Conversion (turning a noun into a verb, or a verb into a noun, etc.)

  network (noun) → *to network* (verb)

  *We **networked** all the PCs in the office.*
  *We created a **network** of all the PCs in the office.*

- Compounding (putting two or more words together)

  hand + held → *handheld*

  *I bought a new **handheld** last week.*

  Compounds can be written as two separate words (**flash card**), as two words joined with a hyphen (**solid-state**), or as one word (**handheld**). Unfortunately, there are no rules, and some compounds even change spelling over time. For example, **web site** began as two words, then became hyphenated (**web-site**) and is now written as one word – **website**. Always check your dictionary or Google if you are not sure.

  In pronunciation, compounds normally have the main stress on the first part, and the secondary stress on the second part, for example **'video ,game**.

4 Describing flash drives

A Listen to a salesperson at his stand at a consumer electronics show describing two flash products to a potential customer. Which product (a or b) is the visitor most interested in?

a The Dragon flash drive

b The Dragon MP4 player

B Listen again and tick (✔) which features the salesperson mentions for each device.

Features	Dragon flash drive	Dragon MP4 player
Back up computer data	☑	☐
Transport files between PCs	☐	☐
Audio and video playback	☐	☐
FM radio tuner	☐	☐
Voice recorder	☐	☐
Games	☐	☐

C Listen again and answer these questions.

1. What is the storage capacity of the Dragon flash drive?
2. How do you connect it to the computer?
3. According to the salesperson, what are the advantages of a USB flash drive over a DVD or an external hard drive?
4. Some portable media players are also known as MP4 players. Why?
5. What is the screen size of the Dragon MP4 player?
6. How long does the battery last?

An MP4 player

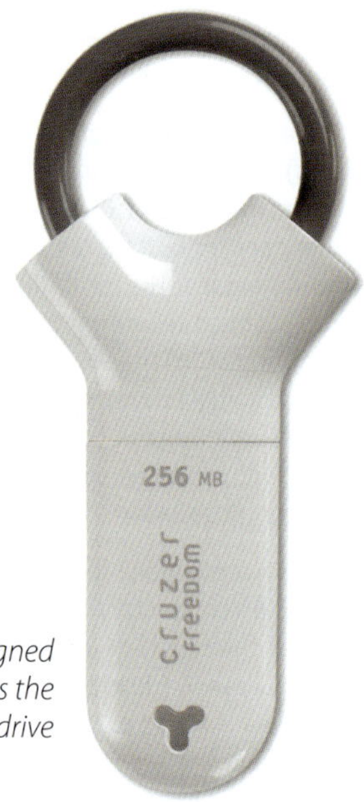

USB drives are typically designed to attach to a key ring, such as the Cruzer Freedom USB flash drive

D **In pairs, choose a flash-based device that you own and describe it. Use the *Useful language* box and the features and questions from the listening text to help you.**

E **You have received a text from a friend at a computer show. Write a short reply.**

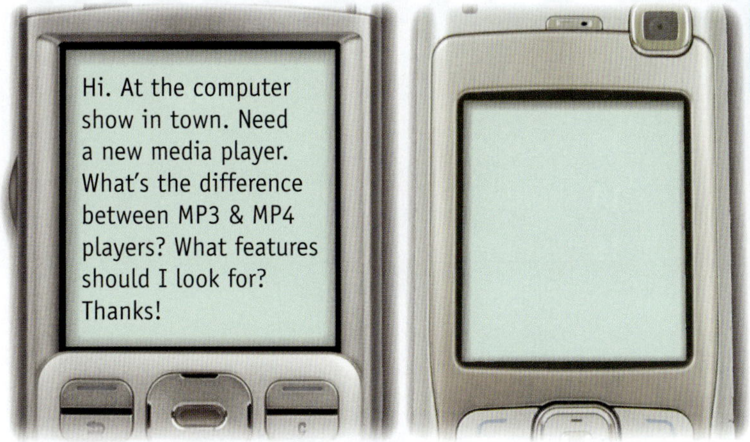

Useful language

It has a storage capacity of …

It features … and …

It supports multiple formats: … and …

You can … and …

Its battery life is …

5 *Vocabulary revision*

Solve the clues and complete the puzzle. Look at Units 10–12 to help you.

Across

4 Thousandth of a second, abbreviated to *ms*, used to measure the access time of hard drives.

6 Floating gate transistors are called .... in flash memory technology.

7 Prefix meaning *very large* or *one thousand million*.

11 Acronym for *light amplification by stimulated emission of radiation*.

12 Capable of being deleted.

Down

1 Concentric ring on the surface of a disc when the disc is formatted.

2 .... memory retains its data when the power is switched off.

3 CD-RW means Compact Disc ....

5 Abbreviation of *digital versatile disc*.

8 To write information on a disk or storage area.

9 Type of external bus or connector that plugs into the computer.

10 The physical mechanism that accepts, reads and writes data on a disk.

Now visit www.cambridge.org/elt/ict for an online task.

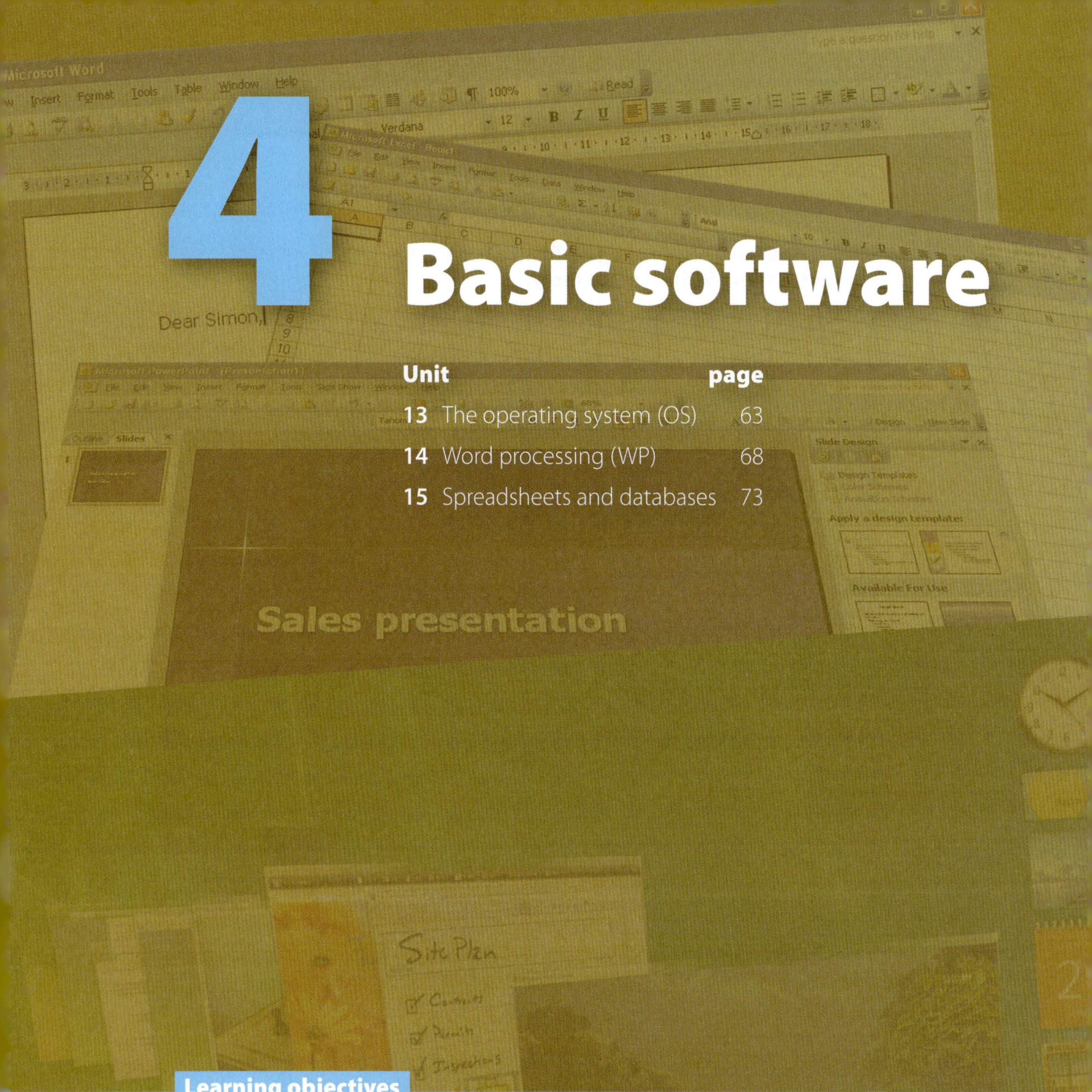

4 Basic software

Unit		page
13	The operating system (OS)	63
14	Word processing (WP)	68
15	Spreadsheets and databases	73

Learning objectives

In this module, you will:

- learn about the function of the operating system.
- learn about the features of a graphical user interface, or GUI.
- practise using the correct determiners with countable and uncountable nouns.
- learn how to summarize a written text.
- learn about the basic features and applications of word processors.
- learn how to give and follow instructions.
- study the basic features and applications of spreadsheets and databases.
- practise forming and pronouncing plurals.

The operating system (OS)

1 The function of the operating system

A In pairs, discuss these questions.

1 How many operating systems can you think of? Make a list.

2 What is the function of an operating system?

B Complete the text with words from the box. If necessary, look at the Glossary.

application software	operating system	software	system software

The set of program instructions that tell the computer what to do is known as (1) .... It can be classified into two basic categories:

- the (2) ...., which includes all the programs that control the basic functions of a computer (e.g. operating systems, programming software, device drivers and utilities).
- the (3) ...., which comprises programs that let you do specific tasks. Typical applications include word processing, databases, educational programs, email and video games.

The (4) .... is a set of programs that control the hardware and software resources of a computer system. Typical functions include handling input/output operations, running programs and organizing files on disks.

2 GUI operating systems

A In pairs, discuss these questions.

1 What does *user-friendly* mean?

2 Do you think most operating systems are user-friendly? Give reasons for your answers.

B Read the text on page 64 and decide which adjectives in the box best describe a GUI.

user-friendly	slow	accessible	text-based	intuitive	complex	graphics-based

C Read the text again and answer these questions.

1 What kind of OS was used in the early 80s: text-based or graphics-based?

2 What is the contribution of Macintosh computers to the development of graphic environments?

3 What does the acronym WIMP stand for?

4 How do you run a program on a computer with a graphical interface?

5 What is *multitasking*?

6 Which multi-user OS is used on large, powerful computers?

7 What is the benefit of using open-source software, for example Linux?

8 Which Microsoft platform is used for pocket PCs, mobiles and portable media centres?

GUI operating systems

The term **user interface** refers to the standard procedures that the user follows in order to interact with a computer. In the late 1970s and early 80s, the way users accessed computer systems was very complex. They had to memorize and type a lot of commands just to see the contents of a disk, to copy files or to respond to a single prompt. In fact, it was only experts who used computers, so there was no need for a user-friendly interface.

In 1984, Apple produced the Macintosh, the first computer with a mouse and a **graphical user interface (GUI)**. Macs were designed with one clear aim: to facilitate interaction with the computer. A few years later, Microsoft launched Windows, another operating system based on graphics and intuitive tools. Nowadays, computers are used by all kinds of people, and as a result there is a growing emphasis on accessibility and user-friendly systems.

A **GUI** makes use of a **WIMP** environment: **w**indows, **i**cons, **m**enus and **p**ointer. The background of the screen is called the **desktop**, which contains labelled pictures called **icons**. These icons represent **files** or **folders**. Double-clicking a folder opens a window which contains **programs**, **documents**, or more nested folders. When you are in a folder, you can launch a program or document by double-clicking the icon, or you can drag it to another location. When you run a program, your PC opens a window that lets you work with different tools. All the programs have a high level of consistency, with similar toolbars, menu bars, buttons and dialog boxes. A modern OS also provides access to networks and allows multitasking, which means you can run several programs – and do various tasks – at the same time.

The most popular operating systems are:

- The **Windows** family – designed by Microsoft and used on most PCs. The most recent version is Windows Vista.
- **Mac OS** – created by Apple and used on Macintosh computers.
- **Unix** – a multi-user system, found on mainframes and workstations in corporate installations.
- **Linux** – open-source software developed under the GNU General Public License. This means anybody can copy its source code, change it and distribute it. It is used in computers, appliances and small devices.
- **Windows Mobile** – used on most PDAs and smartphones (PDAs incorporating mobile phones).
- **Palm OS** – used on Palm handheld devices.
- **RIM** – used on BlackBerry communication devices. Developed by Research In Motion.
- The **Symbian OS** – used by some phone makers, including Nokia and Siemens.

These computer platforms differ in areas such as device installation, network connectivity or compatibility with application software.

D Translate these terms and expressions into your own language. Use a dictionary or the Internet to help you.

1 user interface (line 1) ....
2 procedures (line 2) ....
3 commands (line 6) ....
4 tools (line 16) ....
5 desktop (line 21) ....
6 nested folders (line 25) ....
7 launch a program (line 26) ....
8 source code (line 45) ....

E Label the interface features (a–j) on the screenshot of Apple's Mac OS X operating system with words in bold from this list.

- **desktop:** the background screen that displays icons and folders
- **window:** a scrollable viewing area on screen; it can contain files or folders
- **icon:** a picture representing an object; for example, a **document**, **program**, **folder** or **hard drive icon**
- **folder:** a directory that holds data, programs and other folders
- **menu bar:** a row of words that open up menus when selected
- **drop-down (pull-down) menu:** a list of options that appears below a menu item when selected
- **scroll bar:** a horizontal or vertical bar that is clicked and dragged in the desired direction
- **dock:** set of icons at the bottom of the screen that give you access to the things you use most

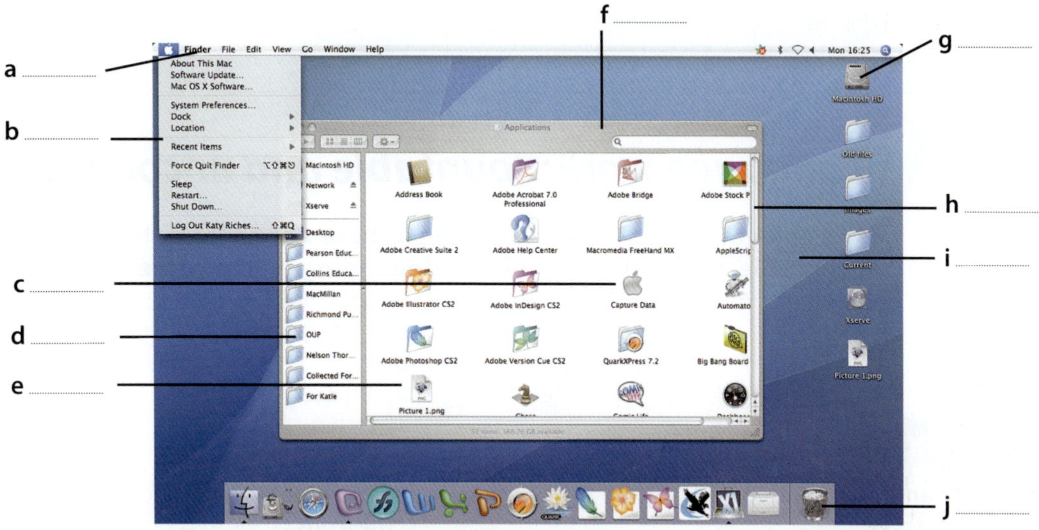

F Compare the Mac OS X user interface with a Windows or Linux interface. What are the similarities and differences? Which features do you prefer from each interface?

3 *Windows Vista*

A Listen to a podcast interview with Bill Thompson, a program developer, and answer these questions.

1 Why is Windows so popular? Give two reasons.

2 Which Windows Vista edition is aimed at high-end PC users, gamers and multimedia professionals?

Windows Vista

B Listen again and complete this fact file.

Windows Vista editions	Other features	Internet and security	Windows programs
(1) .... is designed for users with basic needs, such as email and internet access. Home Premium is for advanced home computing and (2) .... The Business edition is ideal for (3) .... The Ultimate edition is the most complete.	The user interface has been redesigned with new icons and a new (4) .... It offers support for the latest technologies, from DVD creation to (5)	Internet Explorer is more reliable and secure. The Security Centre includes an (6) .... program called Windows Defender, and a firewall that protects your computer from (7)	The most popular is still (8) .... , a suite that includes the (9) .... , Word; an email program; the Excel spreadsheet program; and the (10) .... program, PowerPoint.

4 Language work: countable and uncountable nouns

A **Look at the HELP box and decide if these nouns from the fact file in 3B are countable, uncountable or either, depending on the context. Write *C, U,* or *C and U*.**

user .... email .... computing ....
edition .... entertainment .... interface ....
icon .... technology .... security .... spyware ....

HELP box

Countable and uncountable nouns

- Countable nouns are people or things that we can count. They have a singular and a plural form (e.g. **file**, **program**, **system**, **application**).
- Uncountable nouns are things that we can't count. They have no plural form (e.g. **software**, **music**, **robotics**, **multimedia**, **networking**, **storage**).

  *A lot of **software** these days is open-source.*
  Not: *A lot of ~~**softwares**~~ these days ~~**are**~~ open-source.*
- Some words are countable in many languages but uncountable in English, and are used with a singular verb (e.g. **advice**, **damage**, **equipment**, **furniture**, **research**, **news**, **progress**, **homework**).

  *The **advice** he gave me **was** very useful.*
- Countable nouns must have a determiner (**a**, **the**, **my**, **this**, etc.) in the singular, although this is not necessary in the plural.

  *I deleted **the file** yesterday.*
  *I lost more than 300 **files** when my computer crashed.*

  We use **a** before a consonant sound and **an** before a vowel. The definite article **the** means *you know which one/ones I mean.*

  ***An icon** is a small graphic.*
  ***The icons** on the toolbar are used to ...*
- We don't use **a/an** with uncountable nouns.

  Not: *~~a robotics~~*
- We don't use **the** in generalizations with uncountable nouns or plural countable nouns.

  *I like **music**.*
  Not: *I like ~~**the**~~ **music**.*
  ***Computer programs** are expensive.*
  Not: *~~**The**~~ **computer programs** are expensive.*
- Countable and uncountable nouns take different determiners.

  Many, **few**, **a few** only go with countable nouns.

  *There are **many versions** of Windows Vista.*

  Much, **little**, **a little**, **a great deal of** only go with uncountable nouns.

  *I have **a little time** free this afternoon if you want to meet.*

B Complete this text with *a, an, the* or nothing.

Linux is (1) .... operating system and it was initially created as (2) .... hobby by a young student, Linus Torvalds, at the University of Helsinki in Finland. Version 1.0 of the Linux Kernel* was released in 1994. (3) .... Kernel, at the heart of all Linux systems, is developed and released under GNU General Public License, and its source code is freely available to everyone.

Apart from the fact that it's freely distributed, (4) .... Linux's functionality, adaptability and robustness has made it the main alternative for proprietary Unix and Microsoft operating systems. IBM, Hewlett-Packard and other giants of the computing world have embraced Linux and support its ongoing development. More than (5) .... decade after its initial release, Linux is being adopted worldwide, primarily as (6) .... server platform. Its use as a home and office desktop operating system is also on the rise. The operating system can also be incorporated directly into (7) .... microchips in a process called (8) .... embedding, and it is increasingly being used this way in appliances and devices.

*The Kernel provides a way for software and other parts of the OS to communicate with hardware.

5 *Writing a summary*

Summarize the text on page 64 in 90–100 words. Follow these steps:

1 Read the text again.
2 Underline the relevant information in each paragraph.
3 Make notes about the main points. Leave out details such as examples.
4 Make sentences from the notes and link the sentences with connectors (*and, but, because, therefore*, etc.).
5 Write your first draft.
6 Improve your first draft by reducing sentences. For example:
   - Cut out unnecessary phrases
     Macs were designed ~~with one clear aim:~~ to facilitate interaction with the computer.
   - Omit qualifying words (adjectives or modifying adverbs)
     ~~very~~ complex
   - Transform relative clauses into *-ing* participle clauses
     Double-clicking a folder opens a window which contains programs, documents or ...
     *Double-clicking a folder opens a window **containing** programs, documents or ...*
7 Write the final version of your summary. Don't forget to check the spelling and grammar.

1 Word processing features

A In pairs, discuss these questions.

1 What is a word processor?

2 What kind of tasks do people use word processors for?

3 How many different word processing programs can you name? Which do you think is the most popular?

B Look at this screenshot from Microsoft Word and translate the labelled features and functions into your own language.

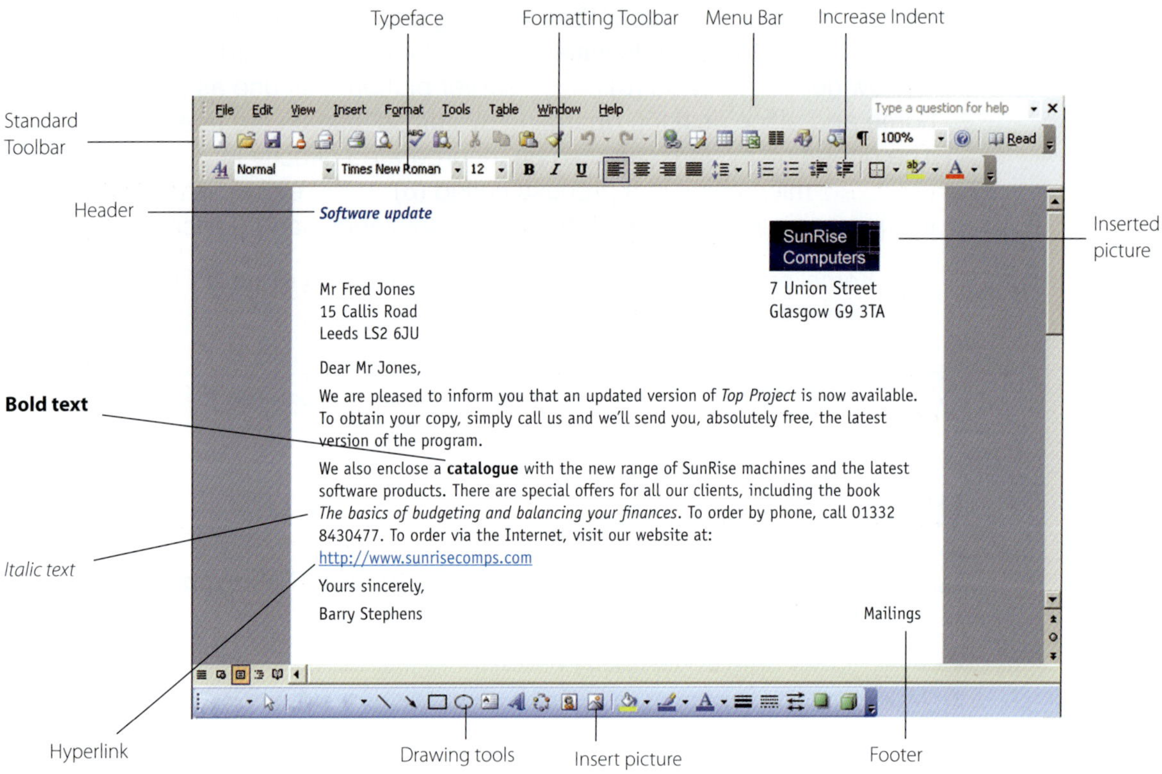

C Complete these sentences with the correct features and functions above.

1 The Standard .... lists the icons to save or print a document, spell check, etc. The .... Toolbar is the area for changing font, alignment, indentation, etc.

2 A font consists of three elements: ...., type style and type size. For example, Palatino bold at 10 points.

3 Type style refers to a visual characteristic of a typeface, for example *B* for ...., *I* for .... and *U* for underlined.

4 If you need to change indentation – the space between the page margin and where the text aligns – you can click the Increase or Decrease .... buttons.

5 The .... and .... commands allow you to specify customized texts at the top and bottom of every page.

2 *Word Sudoku*

In pairs, read the instructions and complete the puzzle.

Instructions

This Word Sudoku is a variation on the normal Sudoku. Instead of using the numbers 1 to 9, we are using words and icons. There are nine WP functions and their equivalent icons, so we are playing with nine pairs. In order to complete the grid, you can use each function or the equivalent icon only once in each row, each column, and in each of the 3x3 boxes. The icons can only be used in the coloured boxes.

Word processing functions and icons

	Align Left		Insert Hyperlink
	Print Preview		Columns
	Insert Table		Undo
	Drawing		Open
	Bullets		

	Drawing	Columns	Bullets					
Align Left			Insert Table					
		Undo			Print Preview			
	Print Preview			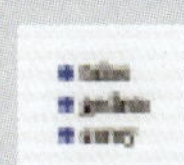		Insert Hyperlink		
								Columns
Undo	Insert Hyperlink	Open	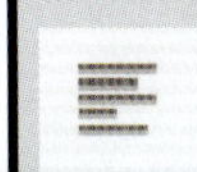					Drawing
						Open	Bullets	
			Columns					Insert Hyperlink
				Insert Table	Insert Hyperlink	Drawing		

3 *The* Cut and Paste *technique*

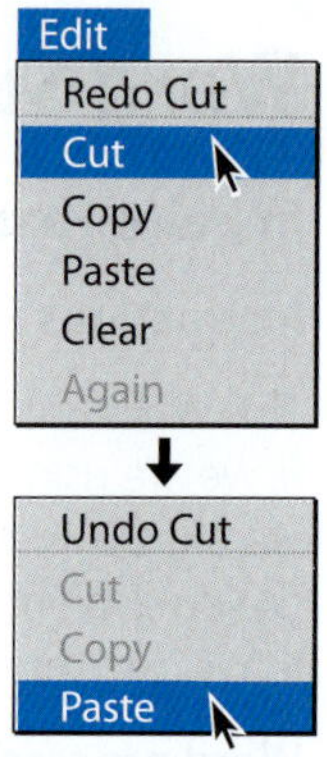

A Listen to two friends, Anna and Ben, talking about how to move text in Word. How many steps are involved in carrying out the *Cut and Paste* task?

B Listen again and complete the dialogue.

Anna: Ben, do you know how I can move this paragraph? I want to put it at the end of this page.

Ben: Er ... I think so. (1) ...., use the mouse to select the text you want to move. (2) .... choose the *Cut* command from the Edit menu.

Anna: (3) .... ?

Ben: Yes. The selected text disappears and goes onto the clipboard. (4) .... you find where you want the text to appear and you click to position the insertion point there.

Anna: Mm, OK. Is that (5) .... ?

Ben: Yes, if that's where you want it. (6) ...., choose *Paste* from the Edit menu, or hold down *Ctrl* and press *V*. (7) ...., check that the text has appeared in the right place.

Anna: OK, I've (8) .... Is that (9) .... ?

Ben: Yes, that's it. If you make a mistake, you can choose *Undo* from the Edit menu, which will reverse your last editing command.

Anna: Brilliant! Thanks a lot.

Ben: That's OK, it's my pleasure.

Moving text is a process of cutting and pasting, as if you were using scissors and glue

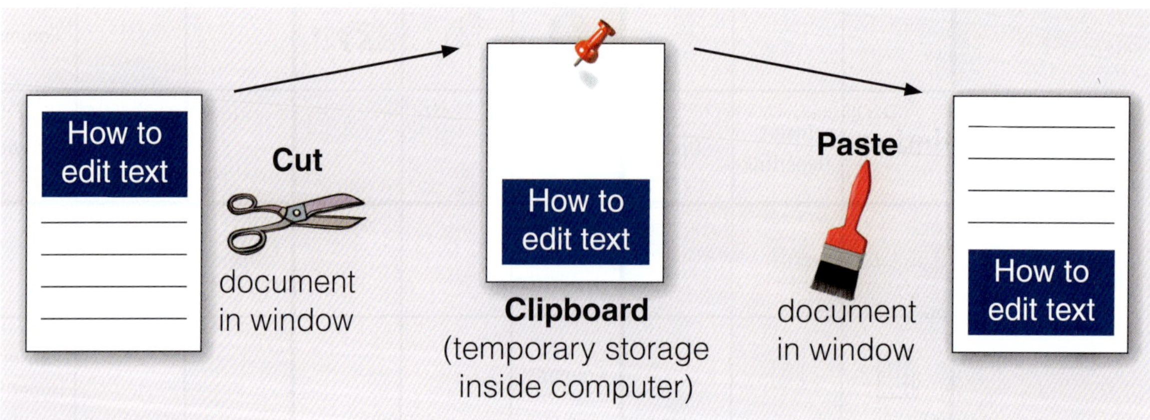

4 Language work: giving and following instructions

A Look at the HELP box and then correct six mistakes in this dialogue.

A: I need a photo for my curriculum vitae. How do I insert one into this Word document?

B: Well, now choose *Insert* on the Menu bar.

A: As this?

B: Yes. From the Insert menu, select *Picture*. As you can see, this displays a drop-down menu with different options: *Clip Art*, *From File*, *From Scanner*, *Chart*, etc. Select *From File* and you'll get a dialog box.

A: OK. I've done that now. What last?

B: OK. Now I navigate your hard drive's contents and find the picture that you want to insert.

A: Right. I'd like to include this one.

B: OK, good. Now click *Insert* and the photograph will be inserted into your document.

A: Here it is. Is that write?

B: Yes. First, right-click with the mouse and select *Format Picture* to adjust the size and other properties.

A: Brilliant, thanks!

HELP box

Giving instructions

- To give instructions, we use the imperative form of the verb and sequence words such as **first**, **next**, **then**, **after that**, **finally**, etc.

  First, use *the mouse to select the text.*
  Then choose *the* Cut *command from the Edit menu.*
  Next, choose Paste *from the Edit menu.*
  Finally, check *that the text has appeared in the right place.*

  We can also use the present simple with **you**.

  *Now **you find** where you want the text to appear and **you click** to position the insertion point.*

Following instructions

- If you want to check that you have understood instructions, you can use expressions like:
  Like this?
  Is that right?
- If you want to signal that you are ready to move on to the next step, you can use expressions like:
  OK, I've done that now.
  What next?
- If you want to ask if the process is completed, you can use expressions like:
  Is that everything?
  Anything else?

B Complete these instructions for how to *Copy and Paste* in Word with verbs from the box.

click (x2) select position right-click drag

1 First, ... the text you wish to copy. To select text, ... the mouse over the portion of the text that you want to copy. This part should then be highlighted.

2 Then .... on the *Copy* icon on the Standard Toolbar. This copies the selected text to an invisible clipboard.

3 Next, ... the cursor where you want the text to appear.

4 Finally, ... the *Paste* icon. This inserts the content of the clipboard at the insertion point. As well as the icons on the toolbar, you can use the keys *Ctrl+ C* for *Copy*, and *Ctrl+V* for *Paste*. These options also come up if you .... the selected text.

C  **Write instructions for using *Find and Replace* based on this dialog box.**

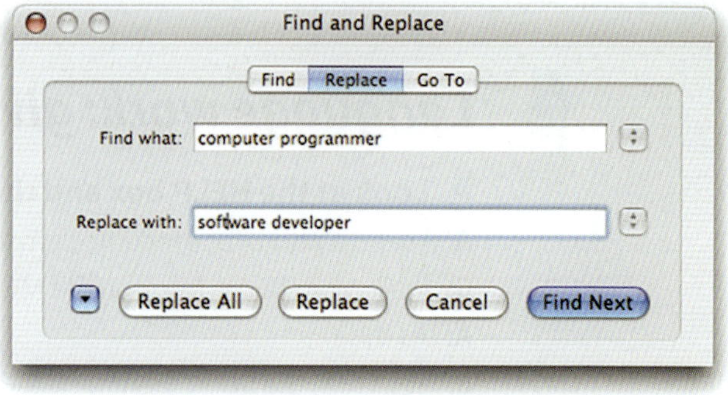

D  **Work in pairs. Student A: Give your partner instructions on *Creating a document and saving it on disk*. Student B: Give your partner instructions on *How to insert a picture from the Web into a Word document*. Use words and expressions from the HELP box on page 71.**

5 *WP tools*

A Scan the descriptions of three WP tools (1–3) – a spell checker, an online thesaurus and a grammar checker – and match them with the dialog boxes (a–c).

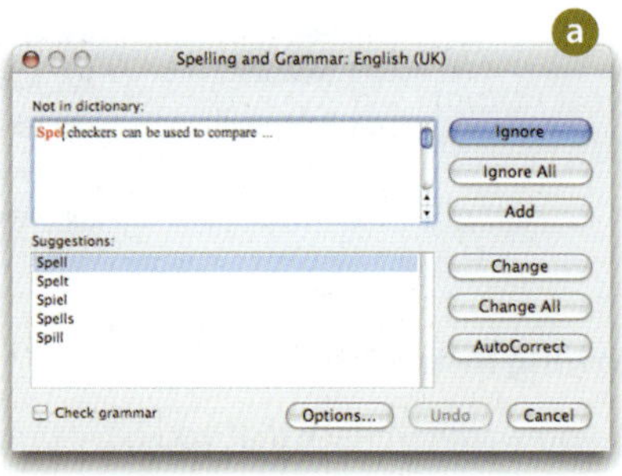

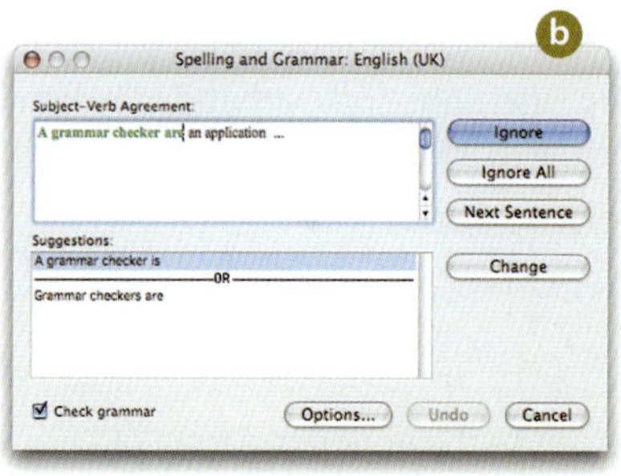

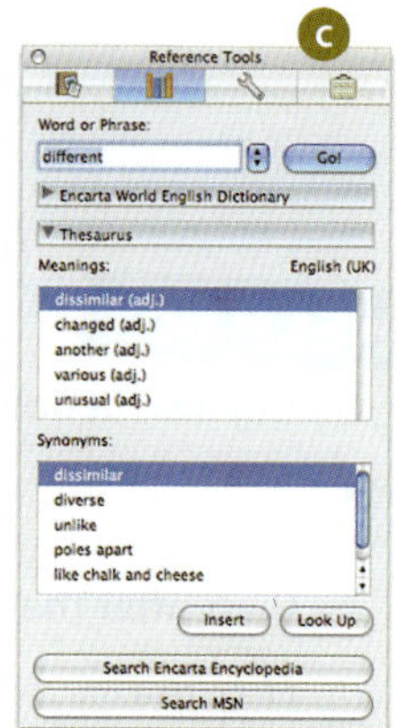

1 Spell checkers can be used to compare words in the program's dictionary to those used in the user's document. The spell checker points out any words it cannot match, notifies the user, and allows them to make any changes; it even suggests possible correct spellings. Like a conventional thesaurus, this database of words contains definitions and suggestions of words with similar and opposite meanings. A word may be spelled correctly but still be wrong (*too* instead of *two*, for instance). This is a good first step at proofing a document because it can find many common errors, but users will still need to proofread documents to ensure complete accuracy.

2 Many word processors include an online thesaurus with which users can look up different words to use in similar instances. Their power comes not from knowing every grammatical rule, but from questioning the writer about certain parts of the text. Some even include information about pronunciation and the history of a word.

3 Grammar checkers are applications that attempt to check more than just spelling. They count words in sentences to flag possible run-on sentences. They look for words that show possible conflicts between verbs and subjects, and they offer advice about corrections. Grammar checkers are a step beyond spell checkers, but they are still not a substitute for a human editor. However, this does not mean that all the words in the document are spelled correctly. They give the writer another chance to think about what he or she has written. The computer can alert writers to problems that wouldn't be obvious to them otherwise.

B Read the descriptions more carefully. Find three sentences that have been printed in the wrong text and decide where they should go.

C Correct the three mistakes in this sentence and decide if they would be found by the spell checker or the grammar checker.

Mail merge combine a form leter with a database file to create customized copys of the letter.

Unit 15 Spreadsheets and databases

1 Spreadsheet programs

A In pairs, discuss these questions.

1 What is a spreadsheet?

2 What are spreadsheets used for?

B Look at the worksheet and label a, b and c with *column*, *row* and *cell*. Then answer these questions.

1 What types of data can be keyed into a cell?

2 What happens if you change the value of a cell?

a .... b ....

c ....

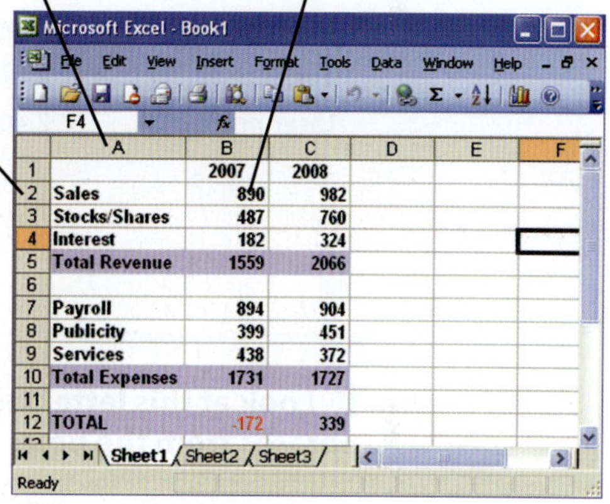

This worksheet shows the income and expenses of a company. Amounts are given in $millions. The terms ***worksheet*** *and* ***spreadsheet*** *are often used interchangeably. However, technically, a* ***worksheet*** *is a collection of cells grouped on a single layer of the file. A* ***spreadsheet*** *refers to both the computer program that displays data in rows and columns, and to the table which displays numbers in rows and columns.*

C Listen to Lucy Boyd giving a training course on basic Excel and check your answers to A and B.

D Listen again and decide whether these sentences are true or false. Correct the false ones.

1 A spreadsheet displays information in the form of a table with a lot of columns and rows.

2 In a spreadsheet you can only enter numbers and formulae.

3 You cannot change the width of columns.

4 Spreadsheet programs can generate a variety of charts and graphs.

5 Spreadsheets cannot be used as databases.

E Look at the worksheet above and decide whether these sentences are true or false. Correct the false ones.

1 The value of the cell C12 is the result of applying the formula *C5-C10*.

2 The value of cell B5 is the result of adding the value in cells B2 and B3.

3 If you type the value *800* in C3, the value in cells C5 and C12 will be recalculated.

F In pairs, discuss the advantages and disadvantages of showing the information above as a graph, rather than as a worksheet.

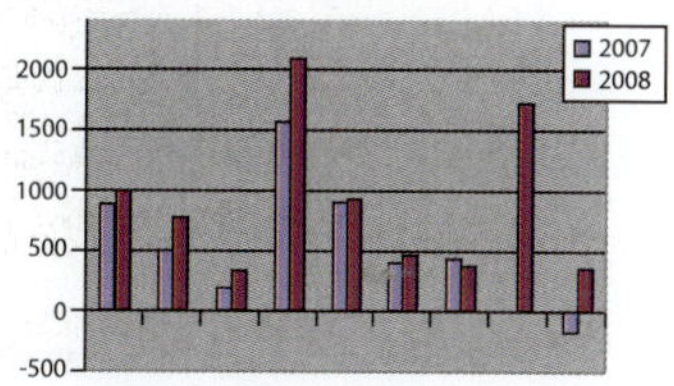

Graphic representation of the worksheet above

2 *An invoice, a business letter and a fax*

A Spreadsheets are also used to generate invoices. Complete the invoice below with words from the box. If you have a spreadsheet program, try to produce a similar invoice.

Quantity | Description | Price | VAT (value added tax) | Product | Grand total | Company

Name: Ruth Atkinson
Address: 38 High Street, Galway
Telephone: 5 742 9165
Date: 16 May 2008

(1) ....
Media Market
Fax: 1 662 2367

(2)	(3)	(4)	(5)	
Ulysses Classic	2GB of RAM, 1TB HD	4	850€	3,400€
Flat LCD screen	Colour 19"	4	170€	680€
Portable Ulysses	2GB of RAM, 250GB HD	2	975€	1,950€
D5 database	DBMS, relational database	1	245€	245€
Antidote JP	Anti-virus, anti-spyware	6	60€	360€
Laser printer CQ	2,400 dpi, PostScript	1	230€	230€
			Sub-total	**6,865€**
		(6)	(21%)	**1,441€**
		(7)		**8,306€**

B Look at this letter which accompanies the invoice. Complete the letter with phrases from the box.

Yours sincerely | I am writing to | Dear Ms Atkinson | We would be grateful if you could | I am enclosing | Please contact us

16 May 2008

Ruth Atkinson
38 High Street
Galway

(1) ....,

(2) .... confirm that we have sent you four desktop PCs plus screens, two laptops and a laser printer, along with a D5 database, and an anti-virus program for each of the computers. Please allow two weeks for delivery.

(3) .... two copies of your invoice.
(4) .... make your payment by cheque or directly to our bank account through the Internet.

We are also delighted to inform you that we are offering our clients an online course called *A paperless office*, free of charge. (5) .... if you require any further information.

(6) ....,

Ian Pegg

C **Imagine you are Ruth Atkinson. When you try to use the laser printer, it gives continuous error messages. You are also having problems installing the database. Write a fax to Media Market to complain. Ask for a new printer and an upgraded version of the database. Look at the *Useful language* box to help you.**

FAX MESSAGE

To: Media Market

Fax: 1 662 2367

From: Ruth Atkinson

Subject: Faulty products

Dear Mr Pegg,

....

....

....

....

Number of pages: 1

Please call if you experience any transmission problems.

Useful language

I am writing to complain about … *… doesn't work* *I am unable to …*

3 *Databases*

A **In groups, make a list of as many possible applications for databases as you can think of.**

Example: *Companies use databases to store information about customers, suppliers and their own personnel.*

B **Look at the illustration, which represents a database file. Can you identify a *record* and a *field*?**

C **Read the text on page 76 and check your answers to B.**

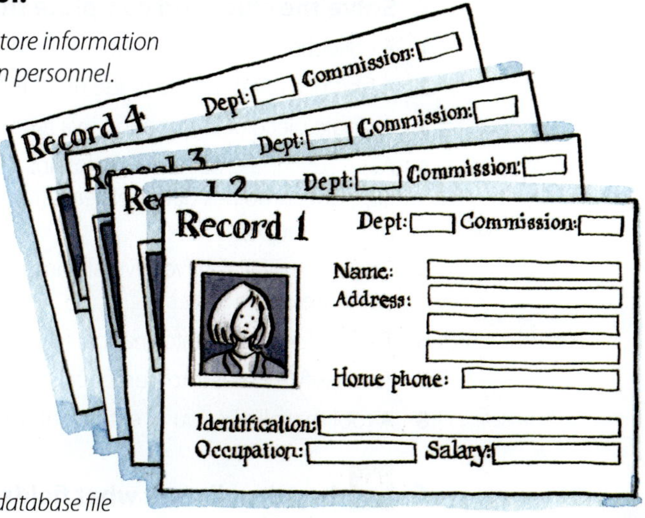

A representation of a database file

Databases

A **database** is a collection of related data, and the software used in databases to store, organize and retrieve the data is called the **database management system**, or **DBMS**. However, we often use the word *database* to cover both meanings. A database can manage any type of data, including text, numbers, images, sound, video and hyperlinks (links to websites).

Information is entered into the database via **fields**. Each field holds a separate piece of information, and the fields are grouped together in **records**. Therefore, a record about an employee might consist of several fields which give their name, address, phone number, date of birth, salary and length of employment with the company.

Records are grouped together into **files** which hold large amounts of information. Files can easily be **updated** – you can always change fields, add new records or delete old ones. An electronic database is much faster to consult and update than a card index system and occupies a lot less space. With the right software, you can keep track of stock, sales, market trends, orders and other information that can help your company stay successful.

A database program lets you create an **index** – a list of records ordered according to the content of certain fields. This helps you to **search** the database and **sort** records into numerical or alphabetical order very quickly. Modern databases are **relational** – that is, they are made up of related files: customers and orders, vendors and purchases, students and tutors, etc. Two database files can be related as long as they have a common field. A file of students, for example, could include a field called *Tutor ID* and another file with details of the tutors could include the same field. This key field can be used to relate the two files. Databases like Oracle, DB2 and MySQL can manage these relationships.

A database **query** function allows you to extract information according to certain conditions or criteria. For example, if a managing director wanted to know all the customers that spend more than €8,000 per month, the program would search on the name field and the money field simultaneously.

The best database packages also include **network** facilities, which can make businesses more productive. For example, managers of different departments can have direct access to a common database. Most aspects of the program can be protected by user-defined passwords and other **security devices**. For example, if you wanted to share an employee's personal details but not their commission, you could protect the commission field.

D Complete these statements about databases using information from the text.

1 A database management system is used to ....
2 Information is entered into a database via ....
3 Each field holds ....
4 *Updating* a file means ....
5 Some advantages of a database program over a manual filing system are: ....
6 Access to a common database over a network can be protected by using ....

E Solve the clues and complete the puzzle.

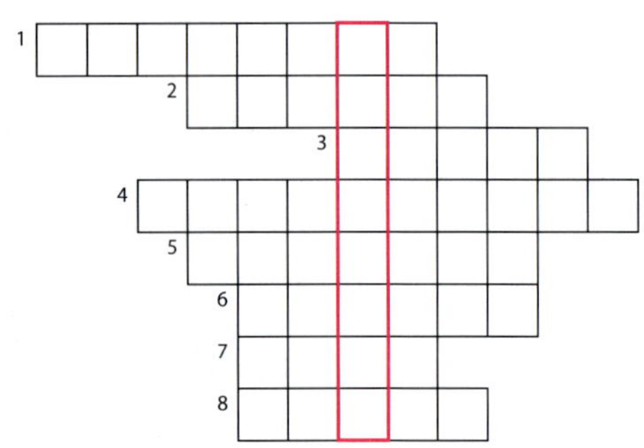

1 A collection of data stored in a PC in a systematic way.
2 A unit of a database file made up of related fields.
3 A single piece of information in a record.
4 A .... database maintains separate, related files, but combines data elements from the files for queries and reports.
5 Some companies have several computers sharing a database over a ....
6 To look for specific information, for example the name of an employee.
7 To classify records into numerical or alphabetical order.
8 A tool that allows you to extract information that meets certain criteria.

F In pairs, discuss what fields you would include in a database for your music collection.

4 Language work: plurals

A Look at the HELP box and then write the plural of these words.

1 client ....
2 key ....
3 query ....
4 businessman ....
5 fax ....
6 salary ....
7 mouse ....
8 virus ....

HELP box

Plurals

- In most cases, we form the plural in English by adding **-s**.

  record → *record**s***

- If a word ends in **-s**, **-sh**, **-x** or **-ch**, we add **-es**.

  address → *address**es***
  index → *index**es***

- If a word ends in a consonant + **y**, the **y** becomes **i** and we add **-es**.

  company → *compan**ies***
  facility → *facilit**ies***

- However, if the **y** follows a vowel, we add only **-s**.

  birthday → *birthday**s***

- There are several irregular plural forms:

  man/woman → ***men/women***
  child → ***children***
  analysis → ***analyses***
  formula → ***formulae*** (or ***formulas***)
  criterion → ***criteria***
  mouse → ***mice***

- The **-s** is pronounced as:

  /s/ after one of these sounds: /p/, /t/, /k/, /f/ or /θ/ (e.g. *amounts, hyperlinks*)

  /ɪz/ after one of these sounds: /s/, /z/, /ʃ/, /tʃ/ or /dʒ/ (e.g. *businesses, devices, images*)

  /z/ in most other cases (e.g. *files, fields, customers, columns*)

B Put the plurals into the correct pronunciation column.

databases	passwords
laptops	graphs
orders	switches
taxes	networks
tables	packages
spreadsheets	systems

/s/	/ɪz/	/z/

5 Software at home and at work

In pairs, find out as much as you can about the software your partner uses at home or at work. Ask about spreadsheet programs, databases, word processors, videoconferencing, business accounting, email, and web browsers. Look at the *Useful language* box to help you.

Now visit www.cambridge.org/elt/ict for an online task.

Useful language

What kind of spreadsheet program do you use?

What do you use it for?

Do you use it at home or at work?

What's your favourite ...?

What features do you like most about it?

How do you ...?

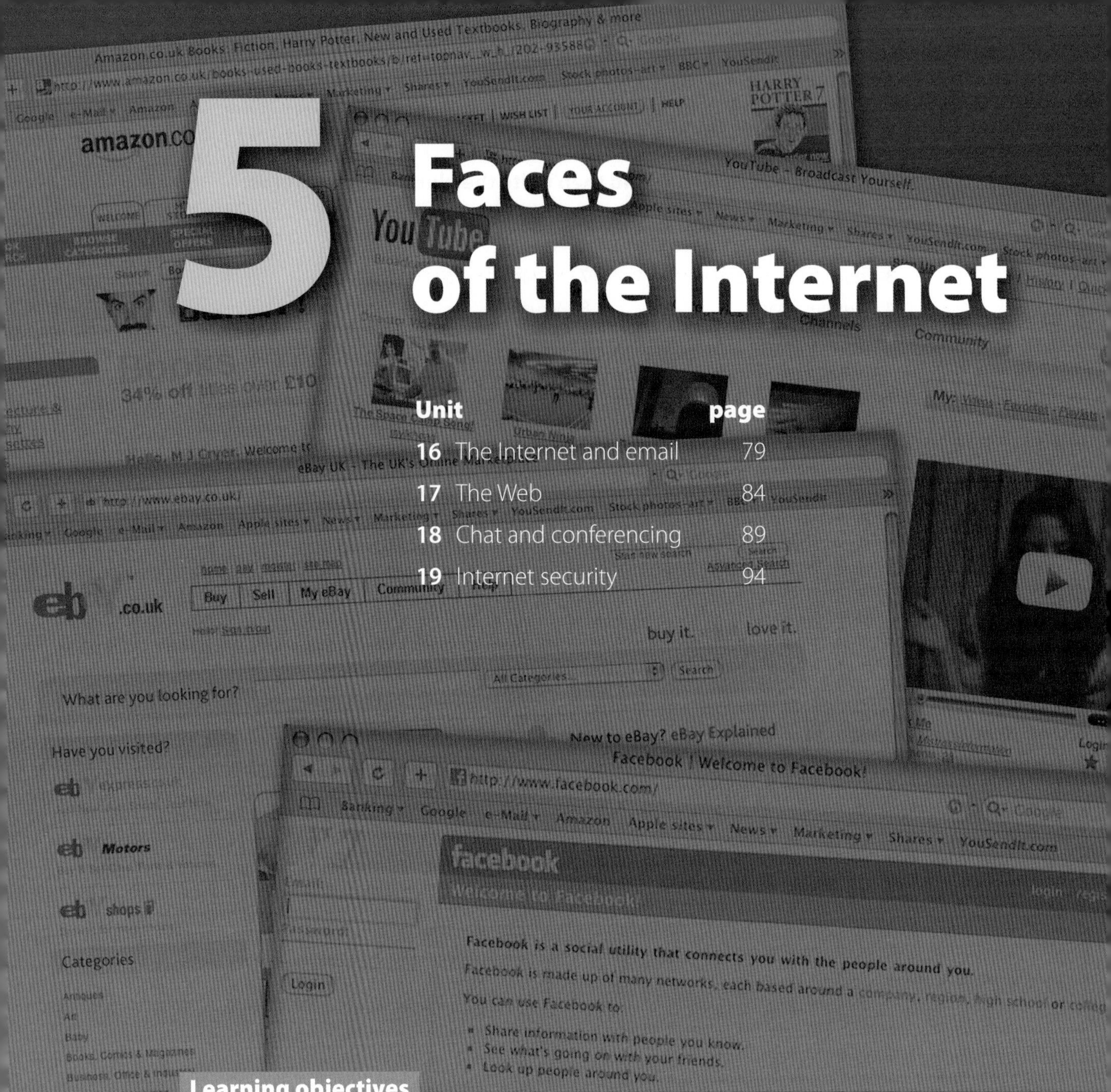

5 Faces of the Internet

Learning objectives

In this module, you will:

- study vocabulary related to the Internet and email.
- learn how to form different types of question.
- learn about the basic features of the Web.
- learn and use collocations related to the Internet.
- learn and use vocabulary related to the Web, e-commerce, online banking, online chatting and videoconferencing.
- learn and use abbreviations in online chats.
- learn about the basic ideas related to security and privacy on the Internet.
- discuss controversial issues related to the Internet.

Unit 16 The Internet and email

1 Internet basics

A **In pairs, discuss how you would define *the Internet*.**

B **Make a list of all the things you can use the Internet for.**

C **Listen to a conversation between a customer buying a PC and a sales assistant. Why do you think the sales assistant has to explain so much about the Internet?**

D **Listen again and complete the customer's notes.**

To connect to the Internet from home, I need:

(1) a .... and (2) a ....

Also need an account with an (3) .... (a company that offers connection for a monthly fee).

If you want to connect lots of computers without using cables, you can use a (4) .... router.

Wi-Fi uses (5) .... waves to send data over medium-range distances.

Things you can do on the Internet:

(6) ....

....

'Web' or 'Internet'? The Web: huge collection of (7) .... stored on computers all over the world. The Internet: the network which connects all the computers.

2 Internet FAQs

A **Read Part 1 of the Internet FAQs on page 80 and choose the correct answers.**

1 The Internet was
   a invented in the mid-90s. **b** popular in the 1960s. **c** probably created in the USA.

2 Which term describes any fast, high-bandwidth connection?
   a broadband **b** dial-up connection **c** Wi-Fi connection

3 The power-line Internet provides broadband access through
   a telephone lines. **b** satellites. **c** electrical power lines.

4 Which device converts computer data into a form that can be transmitted over phone lines?
   a ADSL **b** a mobile phone **c** a modem

5 The standard protocol that allows computers to communicate over the Internet is called
   a an IP address. **b** TCP/IP. **c** HTTP.

6 The geographical region covered by one or several access points is called a
   a wireless access point. **b** hotspot. **c** wireless network device.

Internet FAQs: Part 1

How old is the Internet (the Net)? When was it created?
It's hard to say exactly. The research that led to what we now know as the Internet was begun in the 1960s.

Who created the Internet?
Again, it's hard to say exactly who created it. The initial research was carried out by the Advanced Research Projects Agency in America, funded by the US government.

Did the Internet become popular quickly?
It took many years for the Internet to become popular around the world. It's only really since the mid-90s that the Internet has been a part of our daily lives.

How do you get online?
To get connected, you need a computer, the right connection software and a modem connected to the phone line. You also need an account with an Internet Service Provider (ISP), which acts as a gateway between your PC and the rest of the Net.

How fast are today's internet connections?
Today, ISPs offer a broadband, high-speed connection. The most common types are cable – offered by local cable TV companies – and ADSL (**A**symmetric **D**igital **S**ubscriber **L**ine), which works through phone lines. They are both faster than the traditional dial-up telephone connection. Broadband access is also offered by some electricity networks. This competing technology, known as power-line Internet, provides low-cost access via the power plug, but is still in development.

How long has broadband existed?
Since the late 1990s.

How much does broadband access cost?
It depends on which company you choose. Nowadays, some companies even offer free broadband.

Why do you need a modem?
A modem (**mo**dulator/**dem**odulator) converts digital signals into analogue signals so that data can be transmitted across the phone or cable network.

What does TCP/IP mean?
The language used for data transfer on the Internet is known as TCP/IP (**t**ransmission **c**ontrol **p**rotocol/ **I**nternet **p**rotocol). This is like the internet operating system. Every computer connected to the Net is identified by a unique IP address.

Are there other ways of accessing the Internet?
Other methods of internet access include Wi-Fi, satellite, mobile phones and TV sets equipped with a modem. Wi-Fi-enabled laptops or PDAs allow you to connect to the Net if you are near a wireless access point, in locations called hotspots (for example, a Wi-Fi café, park or campus). Satellite services are used in places where terrestrial access is not available (for example, on ships at sea). High-end mobile phones provide access through the phone network.

B **In pairs, discuss which of the internet systems (1–6) you would use to do the tasks (a–f). Then read Part 2 of the FAQs on page 81and check your answers.**

1 Email
2 The Web
3 Newsgroups
4 Chat and IM
5 FTP
6 Telnet

a transfer files from the Internet to your hard drive
b send a message to another person via the Internet
c have a live conversation (usually typed) online
d connect to a remote computer by entering instructions, and run a program on it
e take part in public discussion areas devoted to specific topics
f download and view documents published on the Internet

Internet FAQs: Part 2

Email

Email lets you exchange messages with people all over the world. Optional attached files can include text, pictures and even audio and animation. A mailing list uses email to communicate messages to all its subscribers – that is, everyone that belongs to the list.

Which email program is the best?
Outlook Express is a popular program, but many users use web-based email accounts such as Hotmail.

The Web

The Web consists of billions of documents living on web servers that use the HTTP protocol. You navigate through the Web using a program called a web browser, which lets you search, view and print web pages.

How often are web pages updated?
It depends entirely on the page. Some are updated thousands of times a day.

Chat and Instant Messaging (IM)

Chat and Instant Messaging technologies allow you to have real-time conversations online, by typing messages at the keyboard.

FTP

FTP, or **f**ile **t**ransfer **p**rotocol, is used to transfer files over a TCP/IP network. Nowadays, this feature is built into Web browsers. You can download programs, games and music files from a remote computer to your hard drive.

Telnet

Telnet is a protocol and a program used to log onto remote computer systems. It enables you to enter commands that will be executed as if you were entering them directly on the remote server.

Newsgroups

Newsgroups are the public discussion areas which make up a system called *Usenet*. The contents are contributed by people who post articles or respond to articles, creating chains of related postings called message threads. You need a newsreader to subscribe to newsgroups and to read and post messages. The newsreader may be a stand-alone program or part of a web browser.

How many newsgroups are there?
There are approximately 30,000 active newsgroups.

Where can you find newsgroups?
Your newsreader may allow you to download the newsgroup addresses that your ISP has included on its news server. An alternative to using a newsreader is to visit web forums instead, which perform the same function but without the additional software.

C Find words and phrases in Part 2 with the following meanings.

1 a system used to distribute email to many different subscribers at once (in *Email* paragraph)
2 a program used for displaying web pages (in *The Web* paragraph)
3 to connect to a computer by typing your username and password (in *Telnet* paragraph)
4 a series of interrelated messages on a given topic (in *Newsgroups* paragraph)
5 a program for reading Usenet newsgroups (in *Newsgroups* paragraph)

3 Language work: questions

A Look at the HELP box and then make a question about Sue Clarke for each of her answers.

1 ....

I'm 23 years old.

2 ....

I'm an online researcher.

3 ....

I use the Internet to find information requested by clients.

4 ....

I've been doing this job for six months.

5 ....

I graduated from university in 2006.

Sue Clarke

HELP box

Questions

- In questions, we normally place the auxiliary verb before the subject.

  Are *there other ways of accessing the Internet?*

- If there is no other auxiliary, we use **do/does** (present simple) or **did** (past simple).

  Did *the Internet become popular quickly?*

- There are many question words in English which we use to find out more information than just *yes* or *no*.

People
Who *created the Internet?*

Things
What *does TCP/IP mean?*
Which *email program is the best?*

Place
Where *can you find newsgroups?*

Time
When *was it created?*
How often *are web pages updated?*
How long *has broadband existed?*

Reason
Why *do you need a modem?*

Quantity
How much *does broadband access cost?*
How many *newsgroups are there?*

Manner
How *do you get online?*

Others
How fast *are today's internet connections?*
How old *is the Internet?*

B In pairs, make questions using these prompts. Then practise asking and answering the questions.

Example: When / first / use the Internet   *When did you first use the Internet?*

1 What type of internet connection / have at home?
2 How fast / your internet connection?
3 How much / pay for broadband access?
4 How often / access the Internet?
5 Which email program / use?
6 Who / send email to?
7 Do / use your mobile phone to access the Internet?
8 Do / use the Internet in public spaces using Wi-Fi?
9 Do / play games online?
10 How many newsgroups / subscribe to?

4 Email features

A Read the text and find the following.

1 the place where your ISP stores your emails
2 the type of program used to read and send email from a computer
3 the part of an email address that identifies the user of the service
4 the line that describes the content of an email
5 the computer file which is sent along with an email message
6 facial symbols used to indicate an emotion or attitude
7 the name given to junk mail

B Write a reply to Celia's email below.

Email features

When you set up an account with an Internet Service Provider, you are given an **email address** and a **password**. The mail you receive is stored on the **mail server** of your ISP – in a simulated mailbox – until you next connect and download it to your hard drive.

There are two ways to get email over the Internet. One is by using a **mail program** (known as an **email client**) installed on your computer, for example Eudora or Outlook Express. The other way is to use **web-based email**, accessible from any web browser. Hotmail and Gmail are good examples.

You can make the message more expressive by including **emoticons**, also called **smileys**. For example, ;-) for wink, :-) for happy, :-o for surprised, :-D for laughing, etc. You may also like to add a **signature file**, a pre-written text file appended to the end of the message. The name given to unsolicited email messages is **spam**.

The anatomy of an email

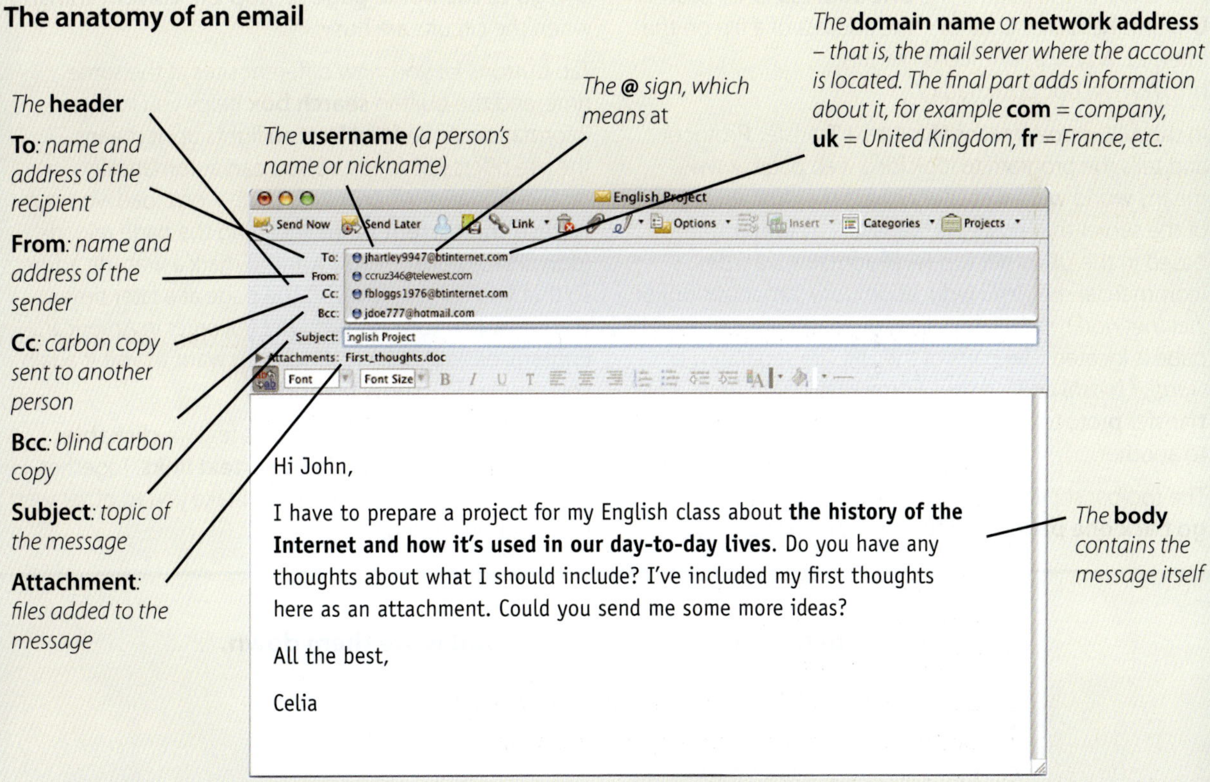

Unit 17 The Web

1 *A typical web page*

A Look at the screenshot of a typical web page. How many of the features (a–k) can you say in English?

a .... b .... c .... d .... e .... f ....

g ....

h ....

i ....

j ....

k ....

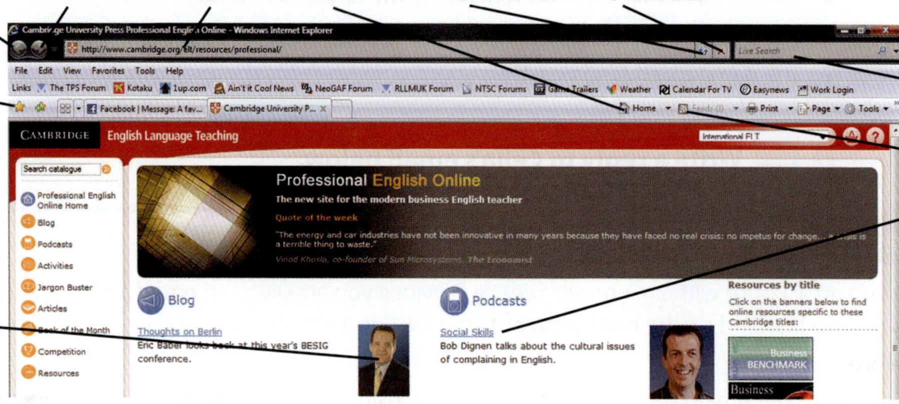

A screenshot from Internet Explorer 7, a leading web browser.

B Read the text and label the features on the screenshot with the terms in bold.

A typical web page

At the top of the page is the **URL address**. URL means **U**niform **R**esource **L**ocator – the address of a file on the Internet. A typical URL looks like this: http://www.bbc.co.uk/radio/.

In this URL, *http://* means **H**yper**t**ext **T**ransfer **P**rotocol and tells the program to look for a web page. *www* means **w**orld **w**ide **w**eb. *bbc.co.uk* is the domain name of the server that hosts the website – a company based in the UK; other top-level domains are *.com* (commercial site), *.edu* (education), *.org* (organization) or *.net* (network); *radio* is the directory path where the web page is located. The parts of the URL are separated by *. (dot)*, */ (slash)* and *: (colon)*. Some sites begin *ftp://*, a **f**ile **t**ransfer **p**rotocol used to copy files from one computer to another.

The toolbar shows all the navigation icons, which let you **go back one page** or **go forward one page**. You can also **go to the home page** or **stop the current transfer** when the circuits are busy.

Tab buttons let you view different sites at the same time, and the built-in **search box** helps you look for information. If the **feed button** lights up, it means the site offers RSS feeds, so you can automatically receive updates. When a web page won't load, you can **refresh the current page**, meaning the page reloads (downloads again). If you want to mark a website address so that you can easily revisit the page at a later time, you can add it to your *favourites* (*favorites* in American English), or bookmark it. When you want to visit it again you simply click **show favourites**.

On the web page itself, most sites feature **clickable image links** and **clickable hypertext links**. Together, these are known as *hyperlinks* and take you to other web pages when clicked.

C Listen to three internet addresses and write them down.

1 ....

2 ....

3 ....

2 The collectives of cyberspace

A Read the article and find websites for the following tasks.

1 to search for information on the Web
2 to buy books and DVDs
3 to participate in political campaigns
4 to view and exchange video clips
5 to manage and share personal photos using tags
6 to buy and sell personal items in online auctions
7 to download music and movies, sometimes illegally

Tour the Collectives of Cyberspace

The Internet isn't just about email or the Web anymore. Increasingly, people online are taking the power of the Internet back into their own hands. They're posting opinions on online journals – weblogs, or blogs; they're organizing political rallies on **MoveOn.org**; they're trading songs on illegal file-sharing networks; they're volunteering articles for the online encyclopedia **Wikipedia**; and they're collaborating with other programmers around the world. It's the emergence of the 'Power of Us'. Thanks to new technologies such as blog software, peer-to-peer networks, open-source software, and wikis, people are getting together to take collective action like never before.

eBay, for instance, wouldn't exist without the 61 million active members who list, sell, and buy millions of items a week. But less obvious is that the whole marketplace runs on the trust created by eBay's unique feedback system, by which buyers and sellers rate each other on how well they carried out their half of each transaction. Pioneer e-tailer **Amazon** encourages all kinds of customer participation in the site – including the ability to sell items alongside its own books, CDs, DVDs and electronic goods. **MySpace** and **Facebook** are the latest phenomena in social networking, attracting millions of unique visitors a month. Many are music fans, who can blog, email friends, upload photos, and generally socialize. There's even a 3-D virtual world entirely built and owned by its residents, called **Second Life**, where real companies have opened shops, and pop stars such as U2 have performed concerts.

Some sites are much more specialized, such as the photo-sharing site **Flickr**. There, people not only share photos but also take the time to attach *tags* to their pictures, which help everyone else find photos of, for example, Florence, Italy. Another successful example of a site based on user-generated content is **YouTube**, which allows users to upload, view and share movie clips and music videos, as well as amateur videoblogs. Another example of the collective power of the Internet is the **Google** search engine. Its mathematical formulas surf the combined judgements of millions of people whose websites link to other sites. When you type *Justin Timberlake* into Google's search box and go to the star's official website, the site is listed first because more people are telling you it's the most relevant Justin Timberlake site – which it probably is.

Skype on the surface looks like software that lets you make free phone calls over the Internet – which it does. But the way it works is extremely clever. By using Skype, you're automatically contributing some of your PC's computing power and Internet connection to route other people's calls. It's an extension of the peer-to-peer network software such as **BitTorrent** that allow you to swap songs – at your own risk if those songs are under copyright. BitTorrent is a protocol for transferring music, films, games and podcasts. A podcast is an audio recording posted online. *Podcasting* derives from the words *iPod* and *broadcasting*. You can find podcasts about almost any topic – sports, music, politics, etc. They are distributed through RSS (Really Simple Syndication) feeds which allow you to receive up-to-date information without having to check the site for updates. BitTorrent breaks the files into small pieces, known as chunks, and distributes them among a large number of users; when you download a *torrent*, you are also uploading it to another user.

Adapted from BusinessWeek online

B Read the article again and match the sentence beginnings (1–5) with the correct endings (a–e).

1 A weblog , or blog, is an electronic journal
2 A peer-to-peer system allows
3 You can use a search engine to find
4 BitTorrent is a peer-to-peer protocol used
5 RSS keeps you constantly informed

a web pages on a particular subject.
b for downloading files over the Internet.
c users to share files on their computers.
d about fresh, new content on your favourite websites.
e that displays in chronological order the postings of one or more people.

C Find words in the article with the following meanings.

1 open-source, editable web pages (lines 5–10) ....
2 the same as *electronic retailer*, or online store (lines 10–15) ....
3 a blog that includes video (lines 25–30) ....
4 a program that allows you to make voice and video calls from a computer (lines 30–35)..........
5 an audio broadcast distributed over the Internet (lines 35–40) ....

D Write a short article (80–120 words) for your school/university/work newsletter about the latest internet phenomena (MySpace, eBay, etc.). Talk about any other sites you think are important or will be important in the future.

3 *Language work: collocations 2*

A Look at the HELP box on page 87 and then match the words on the left (1–6) with the words on the right (a–f) to make collocations. There may be more than one possible answer.

1 online
2 take
3 email
4 upload
5 portable
6 official

a friends
b photos
c action
d website
e encyclopedia
f player

B In pairs, make sentences using the collocations above.

C Find the collocations in these sentences and say what type they are.

1 Once you are online , you can browse the Web, visit chat rooms or send and receive emails.
2 Instant messaging can be a great way to communicate with friends.
3 This software may not be fully compatible with older operating systems.
4 Most webcams plug into a USB port.
5 This highly addictive game will keep you playing for hours.
6 Companies are starting to use virtual reality on their websites.

HELP box

Collocations 2

A collocation is a pair or group of words that are often used together. For example, we say ***make phone calls***, not ***do phone calls***.

Here are some common types of collocation:

- verb + noun (see Unit 1)
  surf the Web ***download music***
- verb + particle
  hack into *a computer* ***log onto*** *a bank account*
- adverb + adjective
  highly sensitive *information*
  freely available *on the Web*
- adjective + noun
  mathematical formulas ***up-to-date information***

The word **online** often collocates with other words and can function as adjective or adverb.

Adjective: *They post opinions on* ***online*** *journals.*

Adverb: *A podcast is an audio recording posted* ***online****.*

4 E-commerce and online banking

A **Listen to two extracts from a monthly podcast called *Money Matters*. What is each speaker talking about?**

Speaker 1 .... Speaker 2 ....

B **Listen again and make notes under these headings.**

Speaker 1	Speaker 2
Things people buy online	Things you can do with online banking
Steps for buying online	Biggest issue with online banking
Precautions	Precautions

C **Complete the extracts with words from the box**

authorization fake internet auction shopping cart browse log in steal

1 Occasionally I also buy things on .... sites such as eBay, where people offer and sell things to the highest bidder.
2 First you enter a site dedicated to e-commerce and .... their products.
3 Then you put the items you want to buy into a virtual .... – a program that lets you select the products and buy with a credit card.
4 You may have to .... with a username and a password ...
5 ... for some transactions, you will be required to use a TAN, a transaction .... number.
6 Be aware of *phishing* – you may receive .... emails claiming to be from your bank and asking for personal information or account details in an attempt to .... your identity.

D **Listen again and check your answers.**

5 Language work: the prefixes e- and cyber-

Look at the HELP box and then complete these sentences.

1 A __________ is an employee who uses his company's internet connection during work hours to chat with friends, play games, etc.

2 An __________ is a postcard sent via the Internet.

3 An __________ is a small magazine or newsletter published online.

4 In a __________ you can use computers with internet access for a fee.

5 Examples of __________ include internet fraud, digital piracy, theft of confidential information, etc.

6 In the future, all elections will be carried out using __________ .

7 You can now sign legal documents online using an __________ .

8 __________ will revolutionise the way we take exams.

9 __________ can be used on some websites instead of real money to make purchases. It reduces the risk of fraud.

10 An __________ is like the paper version, but in digital form.

HELP box

The prefixes *e-* and *cyber-*

- The **e-** prefix means *electronic*, and we add it to activities that take place on computers or online, for example **e-business/e-commerce** – business conducted over the Internet. Other examples include: **e-card**, **e-learning**, **e-zine**, **e-voting**, **e-signature**, **e-assessment**, **e-cash**, **e-book** and **e-pal**.

  There are often spelling variations, with or without a hyphen, so always check your dictionary.

- The **cyber-** prefix comes from *cybernetics*, and we use it to describe things related to computer networks, for example **cybercafé** – an internet café. Other examples include: **cybercrime**, **cyberculture**, **cyberslacker** and **cyberspace**.

6 What do you use the Web for?

In pairs, discuss these questions. Give reasons for your answers.

1 What is your favourite search engine to find information on the Web? Why?

2 Do you download music or video clips from the Web? Do you pay for them?

3 Do you buy things online? Is it better to buy online or go to a shop?

4 Have you ever listened to the radio or watched TV online?

5 Do you use the Web to do school/university assignments or projects? How?

Unit 18 Chat and conferencing

1 Online chatting

In pairs, discuss these questions.

1 What is your favourite way to chat on the Internet?
2 How much time do you spend chatting?
3 Do you give out personal details in chat rooms? Why should you be careful about this?

Windows Live Messenger is one of the world's most popular chat programs

2 Virtual meetings

A Read the text and match the headings (1–5) with the gaps at the start of each paragraph (a–e).

1 Cheap calls over the Internet
2 Virtual worlds and online communities
3 Chat rooms on the Web: join the crowd!
4 Real-time videoconferencing
5 Private chats with IM services

A videoconferencing system combines data, voice and video

Virtual meetings

a ....

Imagine you want to assemble a group of people from around the world for a brainstorming session. **Conferencing** programs such as NetMeeting or CU-SeeMe allow virtual workgroups to communicate via the Internet. To **videoconference**, you'll need a webcam. Participants see each other's faces in small windows on their monitors and hear each other's voices on the computer speakers. You can use just audio, video and audio simultaneously, or the screen-sharing capability to collaborate on documents without audio or video.

b ....

Internet telephony, also known as **VoIP** (**V**oice **o**ver **I**nternet **P**rotocol), almost eliminates long-distance phone charges, allowing you to call nearly anywhere in the world for the price of a local call. If you have flat-rate internet access, you can't beat the price – it's practically free. With internet telephony, you can make a voice call from your computer to another person's computer, landline, or mobile phone. You can download telephony software such as Skype or Net2Phone from the Net, and it's even free!

c ....

People also use more traditional **chat conferencing** or **bulletin board systems** (**BBS**s) to communicate online. Note that during chat sessions, participants type messages to each other rather than communicate by voice. Chat software can be used on the Web with your browser to conduct online chat sessions with other users and can accommodate between 50 and 1,000 users simultaneously. Some companies even use chat conferencing on their websites to facilitate communication with customers.

d ....

Chat rooms can be good venues to meet people and discuss topics of mutual interest. But what if you want to chat privately with a friend, family member or business colleague? Then **Instant Messaging**, or **IM**, is the way to go. Many IM services now offer audio and video capabilities, so if you have a microphone and a webcam, you can chat and see who you're talking to. The four most popular IM services are ICQ and AIM (from AOL), Windows Live Messenger, and Yahoo! Messenger. They all work similarly. First, you enrol in the service by creating a username – which is also your screen name – and a password. Next, you build what is known as a buddy list – a list of people that you want to communicate with. When any of the contacts on your list is online, you can start a private chat with that person.

How do you know who's online? When you launch your IM software, it connects with the service's IM server and logs you on. The server checks your buddy list to see if any of your contacts are also logged on. Your list updates to show who is currently online. By clicking on a name you can send text-based messages to that person. After you type your note and click on the *Send* button, the message travels to the IM server, then immediately forwards to your buddy's computer. This all happens in realtime – instantly.

e ....

You can also chat in incredible **3-D worlds** that are built by other users, for example *Second Life*. In these **virtual reality environments** you can play 3-D games and interact with other users via avatar identities. Avatars are 3-D graphical representations of the participants.

Avatars can run, jump, fly, dance and even enable you to express emotions

Paragraphs a–d adapted from www.learnthenet.com

B Read the text again and answer these questions.

1 Why is videoconferencing so useful for virtual workgroups?

2 What special hardware and software do you need to videoconference?

3 Which technology enables people to make phone calls over the Internet?

4 What is the difference between web chat rooms and Instant Messaging?

5 How do you log on to an IM server?

C Find terms in the text with the following meanings.

1 at a fixed price (lines 15–20) ....

2 a central system that provides information about whether users are online and passes instant messages between them (lines 35–40) ....

3 a friend list or contact list (lines 45–50) ....

4 happening immediately and without delay (lines 55–60) ....

5 artificial reality; a 3-D space generated by the computer (lines 60–65) ....

6 characters used when interacting with people online (lines 60–65) ....

3 Netiquette

A In pairs, do this netiquette quiz. Read about netiquette rules on the Web if necessary.

1 Netiquette, or net etiquette, is a general code of behaviour for communicating online.
- True
- False

2 TYPING IN CAPITALS LETTERS looks like:
- the message is very important.
- you're shouting.

3 What should you avoid doing in chat rooms?
- Being respectful
- Giving out personal or financial information

4 Spamming means
- posting stupid comments in chat rooms.
- posting unsolicited advertising messages.

5 Before asking questions in a chat room or posting messages to forums, you should always
- read the FAQs (Frequently Asked Questions).
- introduce yourself and post a test message.

6 Avoid flame wars. Flames are
- angry responses or offensive comments.
- people who break the rules of netiquette.

7 Keep messages short and to the point, and check spelling and grammar.
- True
- False

B Have you ever experienced bad netiquette? Tell your partner what happened.

4 R u free 4 a chat?

A Rewrite this IM chat, using full forms instead of abbreviations. Then look at the HELP box on page 92 to check your answers.

Abby:	BTW, where r u going for ur holiday? *By the way, where are you going for your holiday?*
Sue:	Girona. Have u been?
Abby:	Yes. I went 2 Girona last summer.
Sue:	Did u have a good time?
Abby:	It's great, IMO. How r u going 2 travel?
Sue:	We're flying.
Abby:	Where r u staying?
Sue:	In a youth hostel.
Abby:	IC. IOW, the cheapest place possible!
Sue:	LOL! Yes. BTW, any recommendations?
Abby:	Let me think. I'll send u a msg ASAP.
Sue:	TIA!
Abby:	Got 2 go. BFN!

B Rewrite this IM chat using abbreviations.

Paulo:	By the way, are you free on Saturday?
Emma:	Sure – it would be good to meet face to face. Shall we go for a coffee?
Paulo:	Good plan. Café Moka makes the best coffee, in my opinion.
Emma:	It's the closest to your house in other words!
Paulo:	Laughing out loud! Yes, you're right! But the coffee really is good.
Emma:	See you at 4?
Paulo:	Great. Bye for now.

C In pairs, practise having an online conversation. Write a short note and give it to your partner. Use abbreviations as necessary. Your partner will write a short response and give it back to you. Continue the conversation and try not to talk. Choose one of these topics.

- Your plans for the weekend
- What you did last night
- Your holiday plans
- What happened at school/work today
- Music / TV / The Web

D In pairs, discuss these questions. Give reasons for your answers.

1 Which program do you use to chat with friends?
2 Do you use abbreviations when you chat online or when you send text messages?
3 Do you use voice or video while chatting? How?
4 Have you ever used the Internet to make cheap calls?
5 Does Instant Messaging distract you from work?
6 Do you use your real name or a nickname in chat rooms?
7 Do you talk to strangers during web chats? Why shouldn't you?
8 Would you ever go on a date with somebody you'd met on the Net?

HELP box

Chat abbreviations

We often use abbreviations in online chats and Instant Messaging. Some common examples are:

ASAP	As soon as possible
BBS	Be back soon
BFN	Bye for now
BTW	By the way
F2F	Face to face
GL	Good luck
H&K	Hug and kiss
IC	I see
ILU	I love you
IMO	In my opinion
IOW	In other words
LOL	Laughing out loud
TIA	Thanks in advance
msg	Message
ur	your/you're
2	to
4	for
b	be
c	see
r	are
u	you

It's OK to use chat abbreviations, but try not to rely on them too much – they can make a conversation difficult to follow. They are also very informal.

www.CartoonStock.com

Computer Dating

5 At a cybercafé

A In pairs, discuss these questions. Give reasons for your answers.

1 Do you ever go to cybercafés?
2 What services would you expect a cybercafé to offer?

B Listen to an interview with Daniel Sturdy, the manager of a cybercafé in London. Does Daniel like where he works?

C Listen again and decide whether these sentences are true or false. Correct the false ones.

1 A cybercafé is a café where you can have access to the Internet and related services.
2 You can talk to people over the Internet using internet telephony at Daniel's café.
3 They don't help people who have problems while using the Internet.
4 Using a computer with internet access costs £2 per hour or £80 for a week.
5 At the moment they've got a lot of international customers.
6 You have to pay long-distance phone rates on the Internet.
7 In the café area you can sit, drink coffee and chat to people.

A cybercafé

6 Plan your own cybercafé

A In small groups, plan how you would open a cybercafé in your town. Consider these areas.

- Money needed
- Type of customer
- Location
- Services you will offer (just internet access? food and drinks? newspapers and magazines? tutorials?)
- Furniture and decoration
- How to create a nice atmosphere (music, lighting, private areas, etc.)
- What type of hardware and software you need
- What type of internet connection you need
- How much you will charge
- A name and slogan for your cybercafé

B Present your plan to the class, using PowerPoint if possible.

Unit 19 Internet security

1 *On alert*

A **In pairs, discuss these questions.**

1 What is a hacker?

2 How easy do you think it is to infiltrate the Internet and steal sensitive information?

3 How can you protect your computer from viruses and spyware?

B **Match the captions (1–4) with the pictures (a–d).**

1 A secure website can be recognized in two ways: the address bar shows the letters *https* and a closed padlock or key is displayed at the bottom of the screen.

2 You have to type your username and password to access a locked computer system.

3 This program displays a message when it detects spyware and other unwanted software that may compromise your privacy or damage your computer.

4 Private networks use a software and/or hardware mechanism called a firewall to block unauthorized traffic from the Internet.

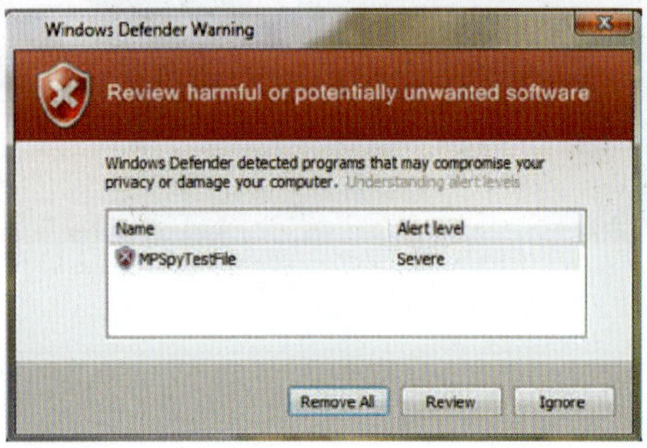

a

b

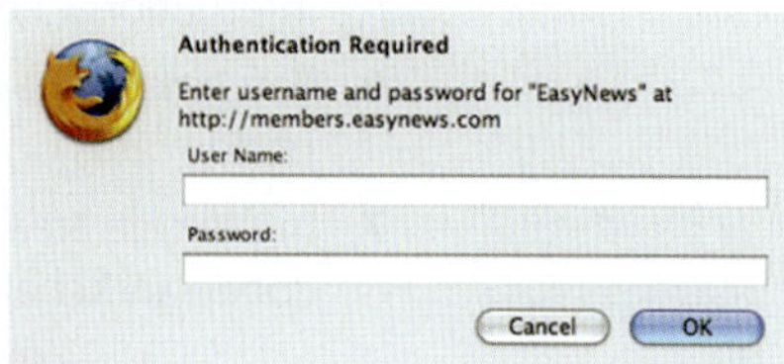

c

d

2 *Security and privacy on the Internet*

A **Read the text quickly and see how many of your ideas from 1A Question 3 are mentioned.**

B **Read the text more carefully and answer these questions.**

1 Why is security so important on the Internet?

2 What security features are offered by Mozilla Firefox?

3 What security protocol is used by banks to make online transactions secure?

4 How can we protect our email and keep it private?

5 What methods are used by companies to make internal networks secure?

6 In what ways can a virus enter a computer system?

7 How does a worm spread itself?

Security and privacy on the Internet

There are many benefits from an open system like the Internet, but one of the risks is that we are often exposed to **hackers**, who break into computer systems just for fun, to steal information, or to spread viruses (see note below). So how do we go about making our online transactions secure?

Security on the Web

Security is crucial when you send confidential information online. Consider, for example, the process of buying a book on the Web. You have to type your credit card number into an order form which passes from computer to computer on its way to the online bookstore. If one of the intermediary computers is infiltrated by hackers, your data can be copied.

To avoid risks, you should set all security alerts to high on your web browser. Mozilla Firefox displays a lock when the website is secure and allows you to disable or delete **cookies** – small files placed on your hard drive by web servers so that they can recognize your PC when you return to their site.

If you use online banking services, make sure they use **digital certificates** – files that are like digital identification cards and that identify users and web servers. Also be sure to use a browser that is compliant with **SSL** (**S**ecure **S**ockets **L**ayer), a protocol which provides secure transactions.

Email privacy

Similarly, as your email travels across the Net, it is copied temporarily onto many computers in between. This means that it can be read by people who illegally enter computer systems.

The only way to protect a message is to put it in a sort of virtual envelope – that is, to encode it with some form of **encryption.** A system designed to send email privately is Pretty Good Privacy, a **freeware** program written by Phil Zimmerman.

Network security

Private networks can be attacked by intruders who attempt to obtain information such as Social Security numbers, bank accounts or research and business reports. To protect crucial data, companies hire security consultants who analyse the risks and provide solutions. The most common methods of protection are **passwords** for access control, **firewalls**, and **encryption** and **decryption** systems. Encryption changes data into a secret code so that only someone with a key can read it. Decryption converts encrypted data back into its original form.

Malware protection

Malware (malicious software) are programs designed to infiltrate or damage your computer, for example **viruses**, **worms**, **Trojans** and **spyware**. A virus can enter a PC via a disc drive – if you insert an infected disc – or via the Internet. A worm is a self-copying program that spreads through email attachments; it replicates itself and sends a copy to everyone in an address book. A Trojan horse is disguised as a useful program; it may affect data security. Spyware collects information from your PC without your consent. Most spyware and adware (software that allows pop-ups – that is, advertisements that suddenly appear on your screen) is included with 'free' downloads.

If you want to protect your PC, don't open email attachments from strangers and take care when downloading files from the Web. Remember to update your **anti-virus software** as often as possible, since new viruses are being created all the time.

Note: Originally, all computer enthusiasts and skilled programmers were known as **hackers**, but during the 1990s, the term hacker became synonymous with **cracker** – a person who uses technology for criminal aims. Nowadays, people often use the word hacker to mean both things. In the computer industry, hackers are known as *white hats* and crackers are called *black hats* or *darkside hackers*.

C Solve the clues and complete the puzzle.

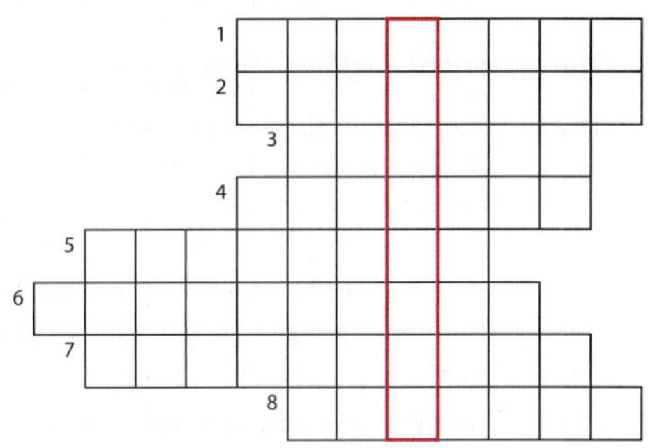

1 Users have to enter a .... to gain access to a network.

2 A .... protects a company intranet from outside attacks.

3 A .... is a person who uses their computer skills to enter computers and networks illegally.

4 .... can infect your files and corrupt your hard drive.

5 You can download .... from the Net; this type of software is available free of charge but protected by copyright.

6 Encoding data so that unauthorized users can't read it is known as ....

7 This company uses .... techniques to decode (or decipher) secret data.

8 Most .... is designed to obtain personal information without the user's permission.

3 Safety online for children

A Listen to an interview with Diana Wilson, a member of the Internet Safety Foundation. Which answers (a or b) best describe what she says?

1 Parents should make children aware of
   - **a** the benefits and risks of the Internet.
   - **b** the risks of the Internet.

2 A web filter program can be used to
   - **a** prevent access to sites with inappropriate content.
   - **b** rate web content with labels (similar to the way movies are rated).

3 If kids spend too much time online or suffer from internet addiction, parents should
   - **a** stop them using the Internet.
   - **b** look for help from specialists.

B Listen again and complete the interviewer's notes.

Risks

Manipulation of children

Invasions of (1) ....

Distribution of indecent or (2) .... material

Violence and racist (3) ....

Solutions

There are websites (4) .... at children.

Internet (5) .... programs let parents block objectionable websites.

Websites should (6) .... their content with a label, from child-friendly to over18 only.

4 The history of hacking

A Read Part 1 of the text and answer these questions.

1 Which hacking case inspired the film *War Games*?

2 When did *Captain Zap* hack into the Pentagon?

3 Why was Nicholas Whitely arrested in 1988?

4 How old was the hacker that broke into the US defence computer in 1989?

The history of hacking – Part 1

1971 – John Draper discovered that a whistle offered in boxes of Cap'n Crunch breakfast cereal perfectly generated the 2,600Hz signal used by the AT&T phone company. He started to make free calls. He was arrested in 1972 but wasn't sent to prison.

1974 – Kevin Mitnick, a legend among hackers, began hacking into banking networks and altering the credit reports of his enemies. He didn't expect that his most famous exploit – hacking into the North American Defense Command in Colorado Springs – would inspire the film *War Games* in 1983.

1981 – Ian Murphy, a 23-year-old known as *Captain Zap* on the networks, hacked into the White House and the Pentagon.

1987 – The IBM international network was paralysed by a hacker's Christmas message.

1988 – The Union Bank of Switzerland almost lost £32 million to hackers. Nicholas Whitely was arrested in connection with virus spreading.

1989 – A fifteen-year-old hacker cracked the US defence computer.

1991 – Kevin Poulsen, known as *Dark Dante* on the networks, was accused of stealing military files.

B **In pairs, discuss which of the cases in Part 1 you had heard of. Which do you think is the most important?**

5 *Language work: the past simple*

A **Look at the HELP box and then complete Part 2 of the text with the past simple form of the verbs in the box.**

show	spread	steal	launch	attempt	overwrite	be	infect	affect

The history of hacking – Part 2

1992 – David L Smith (1) .... prosecuted for writing the Melissa virus, which was passed in Word files sent via email.

1997 – The German Chaos Computer Club (2) .... on TV how to obtain money from bank accounts.

2000 – A Russian hacker (3) .... to extort $100,000 from online music retailer CD Universe.

A Canadian hacker (4) .... a massive *denial of service* attack against websites like Yahoo! and Amazon.

The *ILoveYou* virus, cleverly disguised as a love letter, (5) .... so quickly that email had to be shut down in many companies. The worm (6) .... image and sound files with a copy of itself.

2001 – The *Code Red* worm (7) .... tens of thousands of machines.

2006 – Hackers (8) .... the credit card details of almost 20,000 AT&T online customers. However, subscribers to its service (9) (not) ....

HELP box

Past simple

- We use the past simple to talk about a complete action or event which happened at a specific time in the past.

  Past ———————————————— Now

  *He **began** hacking in 1974.*

- We form the past simple of regular verbs by adding **-(e)d** to the infinitive.

  *John Draper **discovered** that a whistle ...*

  We form questions and negatives using **did/didn't**.

  *When **did** Captain Zap **hack** into the Pentagon?*
  *He **didn't expect** that his most famous exploit ...*

- There are many verbs which are irregular in the past simple.

  *Kevin Mitnick **began** hacking into ...*

  For a list of irregular verbs, see page 166.

  We form questions and negatives for irregular verbs in the same way as for regular verbs. The exception is **be** (see below).

  *When **did** Kevin Mitnick **begin** hacking into ...?*
  *He **didn't begin** hacking until 1974.*

- We form the past passive with the past simple of **be** + the past participle.

  *IBM international **was paralysed** by hackers.*
  *He **wasn't sent** to prison.*
  *Why **was** Nicholas Whitely **arrested** in 1998?*

B Read these landmarks in the history of the Internet and prepare at least five questions in the past simple.

Example: *What happened in 1969? What did Ray Tomlinson do in 1971?*

1969 – The US Defense Department establishes ARPANET, a network connecting research centres.

1971 – Ray Tomlinson of BBN invents an email program to send messages across a network. The @ sign is chosen for its *at* meaning.

1981 – IBM sells the first IBM PC. BITNET provides email and file transfers to universities.

1982 – TCP/IP is adopted as the standard language of the Internet.

1988 – Jarkko Oikarinen develops the system known as Internet Relay Chat (IRC).

1991 – CERN (*Conseil Européen pour la Recherche Nucléaire*) creates the World Wide Web.

1998 – The Internet 2 network is born. It can handle data and video at high speed but is not a public network.

1999 – Online banking, e-commerce and MP3 music become popular.

2001 – Napster, whose software allows users to share downloaded music, maintains that it does not perpetrate or encourage music piracy. However, a judge rules that Napster's technology is an infringement of music copyright.

2004 – Network Solutions begins offering 100-year domain registration.

2006 – Americans spend over $100 billion shopping online.

C In pairs, ask and answer your questions.

6 *Internet issues*

A In small groups, look at the list of cybercrimes and discuss these questions.

1 Which crimes are the most dangerous?

2 Is it fair or unfair to pay for the songs, videos, books or articles that you download? Should copyright infringement be allowed online?

3 What measures can be taken by governments to stop cybercrime?

4 Do you think governments have the right to censor material on the Internet?

5 Personal information such as our address, salary, and civil and criminal records is held in databases by marketing companies. Is our privacy in danger?

Cybercrimes

- **Piracy** – the illegal copy and distribution of copyrighted software, games or music files
- **Plagiarism** and **theft of intellectual property** – pretending that someone else's work is your own
- **Spreading of malicious software**
- **Phishing** (**p**assword **h**arvesting f**ishing**) – getting passwords for online bank accounts or credit card numbers by using emails that look like they are from real organizations, but are in fact fake; people believe the message is from their bank and send their security details
- **IP spoofing** – making one computer look like another in order to gain unauthorized access
- **Cyberstalking** – online harassment or abuse, mainly in chat rooms or newsgroups
- **Distribution of indecent or offensive material**

B Write a summary of your discussion on PowerPoint and present it to the rest of the class.

www. **Now visit www.cambridge.org/elt/ict for an online task.**

6 Creative software

Learning objectives

In this module, you will:

- learn and use vocabulary related to graphics software.
- learn how to describe graphics.
- study the basic features and vocabulary related to desktop publishing.
- discuss the pros and cons of e-publishing versus paper publishing.
- write a letter to a newspaper.
- learn about the main components and applications of multimedia systems.
- learn how to use conditional sentences.
- study the basic principles of web page design.
- learn how to use common modal verbs.
- design a mock home page for a college or company.

Unit 20 Graphics and design

1 *Computer graphics*

A **In pairs, look at the computer graphics (a–d) and discuss these questions.**

1 Which of these computer graphics are three-dimensional (3-D)?
2 What are the advantages of creating 3-D images?
3 Which types of professional might use the computer graphics (a–d)?
4 Who else uses computer graphics in their job? How do they use them?

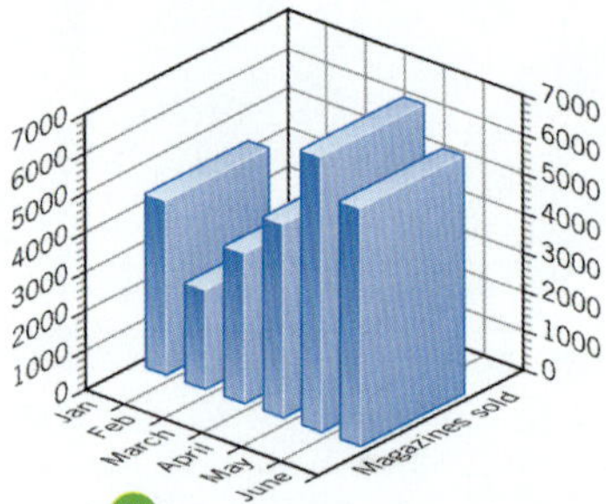

a

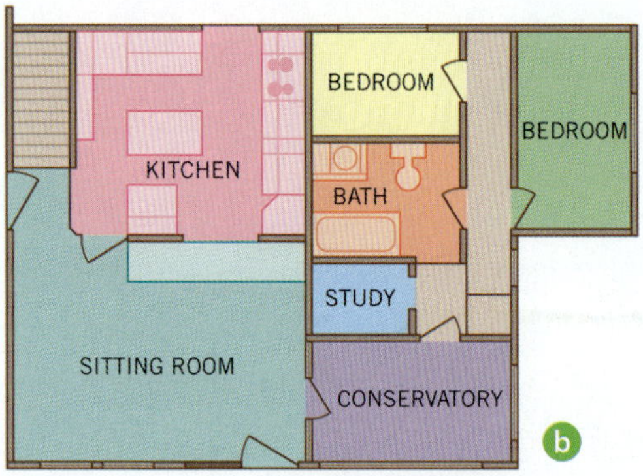

b

c

d

B **Read the text on page 101 and check your answers to 3 and 4 in A.**

C **Read the text again and answer these questions.**

1 What are the differences between *raster* graphics and *vector* graphics?
2 Which graphics file formats are mentioned?
3 What is *compositing*?
4 What does CAD stand for?
5 What are the benefits of using graphics in the car industry?
6 What type of graphics software is used to make maps or 3-D models of the Earth?
7 Who uses computer animation? How?

Computer graphics

Computer graphics are pictures and drawings produced by computer. There are two main categories:

Raster graphics, or **bitmaps**, are stored as a collection of pixels. The sharpness of an image depends on the density of pixels, or **resolution**. For example, text or pictures that are scaled up – that is, made bigger – may show **jagged** edges. Paint and photo-editing programs like Adobe Photoshop focus on the manipulation of bitmaps. Popular raster formats are **JPEG**, **GIF** and **TIFF**.

Vector graphics represent images through the use of geometric objects, such as lines, curves and polygons, based on mathematical equations. They can be changed or scaled without losing quality. Vector data can be handled by drawing programs like Adobe Illustrator, Corel Draw or Macromedia Freehand. **EPS** is the most popular file format for exchanging vector drawings.

◁ *Bitmap graphics are composed of pixels, each of which contains specific colour information*

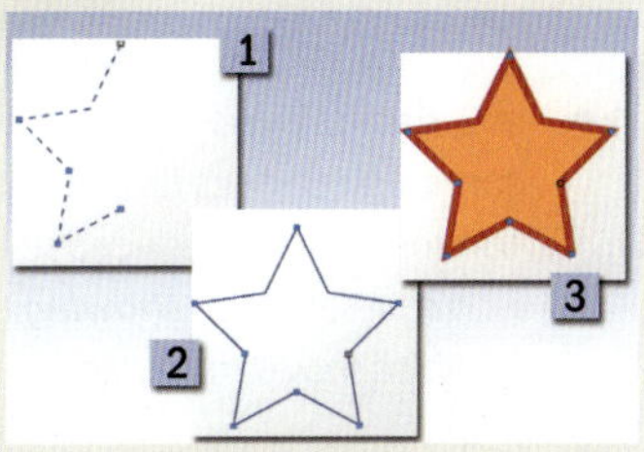

Vector graphics consist of points, lines and curves which, when combined, can form complex objects ▷

Almost all computer users use some form of graphics. Home users and professional artists use image-editing programs to manipulate images. For example, you can add **filters** (special effects) to your favourite photos, or you can **composite** images. Compositing is combining parts of different images to create a single image. Graphic artists and designers use drawing programs to create freehand drawings and illustrations for books or for the Web. Businesspeople use presentation graphics to make information more interesting visually – graphs and diagrams can be more effective ways of communicating with clients than lists of figures. Electrical engineers use graphics to design circuits in order to present data in a more understandable form. Mechanical engineers use **CAD** (**C**omputer **A**ided **D**esign) software to develop, model and test car designs before the actual parts are made. This can save a lot of time and money.

CAD is also used in the aerospace, architecture and industrial sectors to design everything from aeroplanes and buildings to consumer products. Designers start a project by making a **wireframe**, a representation showing the outlines of all edges in a transparent drawing. They then specify and fill the surfaces to give the appearance of a 3-D solid object with volume. This is known as **solid modelling**. Next, they add paint, colour and filters to achieve the desired 'look and feel': this is called **texturing** the object. Finally, they **render** the object to make it look real. Rendering includes lighting and shading as well as effects that simulate shadows and reflections.

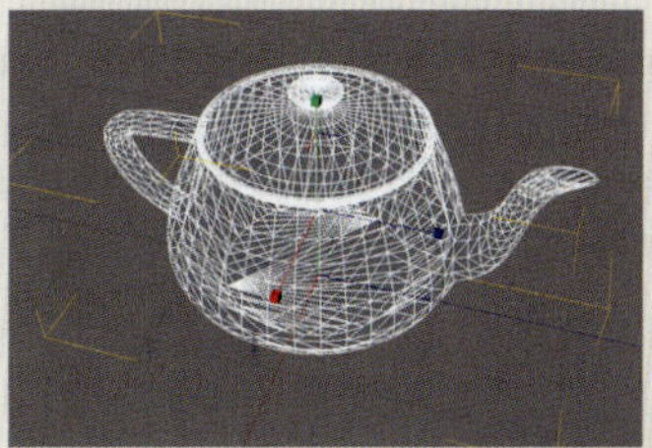

◁ *A wireframe model of a teapot*

Smooth shading – part of the rendering process ▷

Computer art, or **digital art**, is used in adverts and TV programmes. Artists and scientists use special graphic applets to create amazing **fractals**. Fractals are geometrical patterns that are repeated at small scales to generate irregular shapes, some of which describe objects from nature. Government agencies use **GIS** (**Geographic Information Systems**) to understand geographic data and then plan the use of land or predict natural disasters. Cartographers use GIS to make detailed maps. Animators use **computer animation** software to create animated cartoons or add effects in movies and video games.

A fractal

D Match the words (1–6) with the definitions (a–f).

1	resolution	a	special effects that can be applied to pictures
2	jagged	b	a technique that generates realistic reflections, shadows and highlights
3	filters	c	geometrical figures with special properties
4	wireframe	d	irregular or uneven
5	rendering	e	the number of pixels in an image
6	fractals	f	the drawing of a model by using features like edges or contour lines

E In pairs, discuss which application of computer graphics you think is the most important or useful. Give reasons for your answers.

2 *Language work: the -ing form*

A Look at the HELP box and decide if the *-ing* forms in these sentences are gerunds, present participles or adjectives. Write *g, pp* or *a*.

1 PCs generate graphics by performing mathematical calculations on data. ....

2 Businesspeople use graphics to make information more interesting visually. ....

3 Graphs and diagrams can be more effective ways of communicating with clients than lists of figures. ....

4 She is designing a logo for the company. ....

5 If you need to make a presentation, I suggest using PowerPoint. ....

6 The Internet is a network linking other networks. ....

B Correct the mistakes in these sentences. There are seven mistakes in total.

1 Computer animation is the process of create objects which move across the screen.

2 *Texturing* involves add paint, colour and filters to drawings and designs.

3 You can open the colour palette by click on the corresponding icon.

4 CAD programs are very fast at to perform drawing functions.

5 A lot of time and money is saved by test a car design before to make the product.

6 To render refers to the techniques used to make realistic images.

HELP box

The *-ing* form

We use the **-ing** form in three ways:

1 ***Rendering*** *includes* ***lighting*** *and* ***shading****.*
2 *We are* ***designing*** *a new car on computer.*
3 *They use special applets to create* ***amazing*** *fractals.*

- In 1, **rendering** is a gerund (see below), acting as the subject. **Lighting** and **shading** are also gerunds, acting as the objects. A gerund refers to an activity or process.
- In 2, **designing** is a present participle. This is used in continuous tenses (in the above example, the present continuous) and reduced relative clauses.
  ... a representation ***showing*** *the outlines of all edges.*
  (= which shows the outlines ...)
- In 3, **amazing** is an adjective.

We use gerunds in the following ways:

- As the subject of a verb
  Compositing *is combining parts of different images to create a single image.*
- As the complement of the subject
  Compositing is ***combining*** *parts of different images ...*
- As the object of a verb
  I ***enjoy editing*** *pictures.*
- After a preposition
  Designers start a project ***by making*** *a wireframe.*
- As the complement of a verb
  This course ***involves painting*** *and* ***drawing*** *in various media.*
- Some verbs are followed by the gerund, not by the infinitive (e.g. **avoid**, **fancy**, **finish**, **give up**, **hate**, **imagine**, **involve**, **keep**, **look forward to**, **mind**, **suggest**, **enjoy**)

3 The toolbox

A Listen to an extract from an online tutorial about graphics programs and answer these questions.

1 What is a *toolbox* in graphics software?
2 What are graphics *primitives*?
3 What sort of *attributes*, or characteristics, can be used in graphical objects?
4 What does *translation* mean?

B Listen again and complete this extract from the web version of the tutorial.

Graphics programs usually have a *toolbox* – a collection of drawing and (1) .... tools that enable you to type, (2) .... , draw, paint, edit, move, and view images on the computer.

The basic shapes which are used to (3) .... graphical objects are called *primitives*. These are usually geometric, such as lines between two points, arcs, circles, polygons, ellipses and even text. Furthermore, you can specify the *attributes* of each primitive, such as its colour, line type, fill area, interior style and so on.

The various tools in a toolbox usually appear together as pop-up icons in a menu or palette. To use one, you activate it by (4) .... on it. For example, if you want to (5) .... a rectangle, you activate the rectangle tool, and the pop-up options give you the possibility of (6) .... rectangles with square or rounded corners.

You can transform an object by translating, (7) .... or scaling it. *Translation* means moving an object to a different location. *Rotation* is (8) .... the object around an axis. For example, you may need to rotate an object 90 or 180 degrees to fit the drawing. (9) .... is making the object larger or smaller.

C Match the tools from the Photoshop toolbox (1–10) with the functions (a–j).

1 Marquee select tools
2 Move tool
3 Crop tool
4 Paintbrush, pencil
5 Eraser
6 Paint bucket
7 Type tool
8 Colour picker (Eyedropper)
9 Zoom
10 Colour tools and palette

a cut down the dimensions of a picture
b select a particular part of an image (you can choose different shapes for selection)
c fill in an area with a colour
d control the foreground and background colour
e select a specific colour in a photo
f magnify areas of an image when you are doing close, detailed work
g delete the part of the picture you drag it over
h insert text into your document
i draw and paint in different shapes and patterns
j move a selection or entire layer by dragging it with your mouse

4 Choosing graphics software

Work in pairs. Student A chooses a task from the list (1–6) and describes it. Student B chooses the most appropriate graphics software for the task (a–f) and gives reasons for his or her choice. Swap roles. Look at the text on page 101 and the *Useful language* box to help you.

1 to edit and retouch photos
2 to create illustrations and drawings for a magazine
3 to prepare slideshows for training sessions or conferences
4 to make mechanical designs and architectural plans
5 to create dynamic simulations and special effects for films, TV, advertisements and games
6 to analyse geographic data and make maps

a Computer animation software, for example 3-D Studio Max
b GIS software, for example ArcView
c Presentation software, for example PowerPoint
d A CAD package, for example AutoCAD
e Vector graphics software, for example Freehand
f A paint and image-editing program, for example Photoshop

Useful language

If I need to ..., what software would you recommend?

For that kind of task, the best thing would be ...

It allows you to ... and ...

I wouldn't recommend ... because ...

A good program of this type is ...

5 Describing graphics

Look at the images (1–4), which show the stages involved in drawing a plane using computer software. Write a short description of stages 2, 3 and 4. Look at the text on page 101 and the *Useful language* box to help you.

1
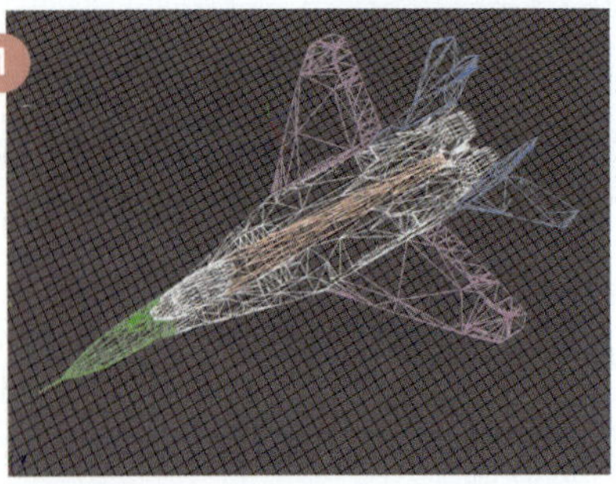
Wireframe

This first image shows a wireframe model, probably made using CAD software. A wireframe is a drawing with edges and contour lines. The parts of the plane are shown in different colours (violet, green, blue, etc.).

Useful language

This picture shows ...

In this (next) stage ...

The designer has used ...

This stage is called ...

Rendering techniques include ...

As a finishing touch, ...

2

Solid modelling

3
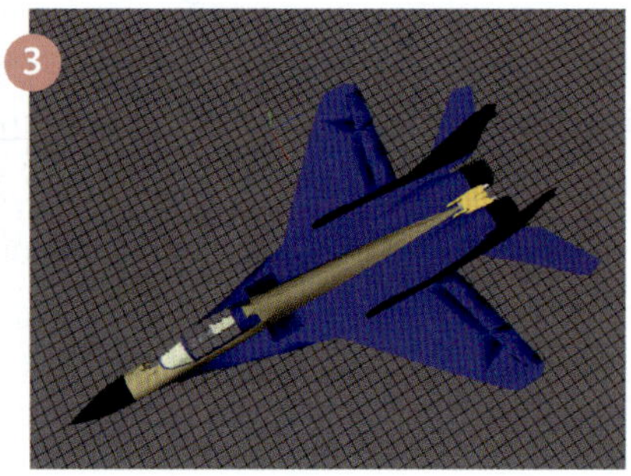
Texturing the model

4

Rendering

Desktop publishing

1 What is desktop publishing?

A **In pairs, discuss these questions.**

1 What kind of documents can be produced with a desktop publishing system?
2 Page layout software is the key component of a desktop publishing system. Which file types can be imported into a page layout program?

B **Read the text and check your answers to A.**

What is desktop publishing?

Desktop publishing (**DTP**) refers to the use of computers to design and publish books, brochures, newsletters, magazines and other printed pieces. DTP is really a combination of several different processes including word processing, graphic design, information design, output and pre-press technologies, and sometimes image manipulation.

DTP centres around a **page layout program**. Typically, a layout program is used to import texts created in word processing programs; charts and graphs from spreadsheet programs; drawings and illustrations created in CAD, drawing or paint programs; and photographs. The program is then used to combine and arrange them all on a page. It is this ability to manipulate so many different items and control how they are used that makes layout software so popular and useful. However, modern word processors also have publishing capabilities, meaning the line separating such programs from DTP software is becoming less clear. In general, though, powerful new publishing systems use high-quality scalable **fonts** and give you control over typographic features such as **kerning** (adjusting the spaces between letters to achieve even, consistent spacing). Another key feature of DTP software is **text flow** – the ability to put text around graphic objects in a variety of ways.

Once composed, DTP documents are printed on a laser printer or on a high-resolution imagesetter (see Unit 8). For transfer to a commercial printer, the documents are generally saved in their native page layout format (such as Adobe InDesign or QuarkXPress) or as **PDF** files. PDF stands for **P**ortable **D**ocument **F**ormat and allows people to view, search and print documents exactly as the publisher intended – you don't need to have the software and fonts used to create it. PDF files can be published and distributed anywhere: in print, attached to email, posted on websites, or on DVD. To open a PDF file, only the Adobe Acrobat Reader (a free download) is required.

In modern commercial printing, DTP files are output directly to the **printing plates** without using film as an intermediate step. This new technology is known as **Computer-To-Plate** (**CTP**) or **direct to plate**, and the machine that generates plates for a printing press is called a **platesetter**. CTP machines are expensive, so most people take their files to a **service bureau**, a company that specializes in printing other people's files. Service bureaux offer a full range of scanning and printing solutions.

C **Read the text again and answer these questions.**

1 What type of software is used for the creation of DTP documents?
2 What are three differences between DTP software and word processors?
3 What is a PDF and what can it do?
4 Which program do you need to view a PDF document?
5 Why do people send their DTP files to service bureaux?

D Find words in the text with the following meanings.

1 shape, style and size of a typeface, for example `Courier at 10pt`
2 the process of adjusting the space between characters
3 feature that enables you to wrap text around images on the page
4 metal surfaces that carry the image to be printed
5 a machine that creates the printing plates

E In pairs, discuss the question *What is desktop publishing?* in as much detail as you can. Then look back at the text on page 105 to see how much you remembered.

2 *Language work: order of adjectives*

A Look at the HELP box and then make phrases using the words in the correct order.

Example: computer programmer / young / clever
a clever, young computer programmer

1 software / desktop publishing / user-friendly
2 hardware company / reliable / young
3 German / industry / graphic design
4 word processing / applications / modern
5 Sony / new / music player / portable

HELP box

Order of adjectives

- Adjectives usually come *before* the noun (also known as the headword).
  They give you control over ***typographic features****.*
  For transfer to a ***commercial printer****, the document is …*
- However, adjectives come *after* certain verbs (e.g. **be**, **look**, **become**, **seem**, **sound**), complementing the subject of the sentence.
  CTP machines ***are expensive****.*
- Adjectives can also complement the object of the sentence.
  This makes layout software ***popular*** *and* ***useful****.*
- This is the usual order of adjectives before a noun:

Opinion	Description	Origin/Place	Material	Purpose	Headword
powerful	new			publishing	systems
high-quality	scalable				fonts
	thin	American	aluminium	printing	plates

Adjectives are ordered from the most subjective (e.g. **nice**) to the most objective (e.g. **silicon**).

Brand names (**Microsoft**, **Sony**, etc.) are considered adjectives of origin/place.

If there is more than one adjective in a sentence, they are usually separated by commas, unless the adjective forms an integral part of the headword (*A fantastic, thin, Sony MP3 player.*)

B Translate these sentences into your own language. How does the use of adjectives differ from English? Think about word order and whether the form of the adjective changes or not.

1 DTP refers to the use of personal computers to produce high-quality printed documents.

2 A page layout application is used to import text from word processing programs and pictures from painting and drawing programs.

3 In modern commercial printing, DTP files are output directly to the printing plates.

C In pairs, choose an object in your classroom or office and think of three words to describe it. Put the words into the correct order and make a sentence.

Example: PC: black, old, DELL

On my desk I've got an old, black, DELL PC.

3 Steps in a DTP publication

A Look at this extract from an online tutorial for DTP publishing. Put the steps in the creation of a DTP document (a–f) into the correct order.

1 [a] 2 [ ] 3 [ ] 4 [ ] 5 [ ] 6 [ ]

a First, the DTP designer decides the basic form of the document (the type of document, general design, colour, fonts, images required, etc.).

b The last step is to take the files to a service bureau, which will print the publication.

c To create the DTP document, the designer begins by selecting a template or by specifying the settings of a new document (the page size, margins, columns, paragraph styles, master pages, etc.).

d When the text has been edited, the designer imports the pictures and uses precise tools to position, scale, crop and rotate all the items.

e The next step is to type the text directly or to import it from a word processing program like Word or WordPerfect.

f Once the file is composed and saved, the designer has to prepare it for printing, which involves verifying the colour specification, creating a Postscript or PDF file, exporting the file in HTML format for the Web, checking proofs, etc.

B Listen to the audio from the online tutorial and check your answers to A.

C Label the features of this page designed with Adobe InDesign (1–6) with words from the box.

toolbox	layout of master pages	dimensions	guide	horizontal ruler	scanned photo

1 ....

2 ....

3 ....

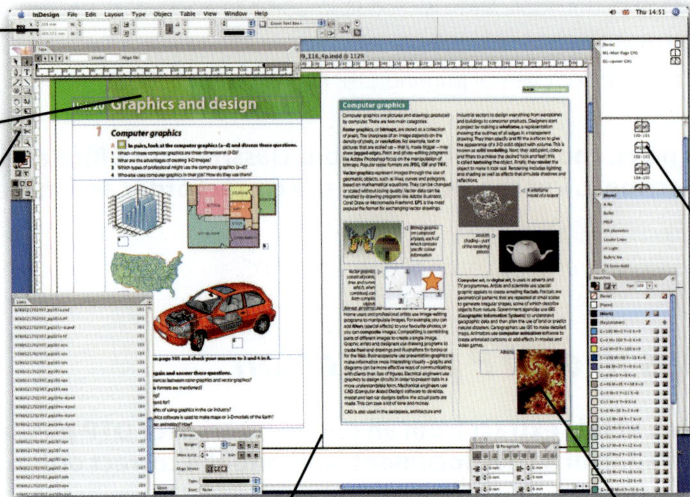

6 ....

4 ....

5 ....

4 *Writing a letter*

A Although most written communication these days is carried out by email, letters are still appropriate for more formal correspondence. Look at this letter. What is the writer asking for?

(a) Rhondda High School
31 Prospect Place, Cardiff, Wales

(b) 28th March 2008

(c) The Editor
El Independiente
Moratin, 7
28006 Madrid
Spain

(d) Dear Sir/Madam,

(e) We are writing to ask if you can help us with our school project. We are doing a survey of the major newspapers in the European Union to find out which computer systems and desktop publishing programs they use.

(f) We would be very grateful if you could tell us which hardware, graphic design and page layout software you use at *El Independiente*. Could you also tell us how long your online edition has been running for? Thank you very much in advance.

(g) We look forward to hearing from you.

(h) Yours faithfully,

Katherine Powell

Katherine Powell, student representative

B Match the parts of the letter (a–h) with the descriptions (1–8).

1 [b] For example, *28th March 2008*, or 28 March 2008, or *28/03/2008*.

2 [ ] This is usually in the top right corner of the letter, but can be in the centre if it's a printed letterhead.

3 [ ] State the reason for writing: *I am / We are writing to … / We are currently …*

4 [ ] This should be included on the left hand side of the page, before the greeting.

5 [ ] Start with *Dear Sir/Madam* or *Dear Mr/Mrs/Ms …* Use *Ms* if you are not sure if the recipient is married or not. It is often best to use *Ms*, as *Mrs* can cause offence.

6 [ ] Make any requests or ask any questions you need to: *We would be grateful if you could …*, Could you also …

7 [ ] Request further contact, if necessary: *We / I look forward to hearing from you. / Please contact us by …*

8 [ ] If you have started the letter with the person's name (for example, *Dear Mr Robinson)*, then end with *Yours sincerely*. If you do not know the name of the recipient, end with *Yours faithfully*.

C Write a letter to a local newspaper, asking for information about the hardware they use in their production, the page layout software they use, and the data communications systems they use. Use A and B above to help you.

5 *E-publishing versus paper publishing*

A Look at this web extract about e-publishing. What examples of e-publishing can you find in the text?

Publishing has existed in its current form for centuries. Ever since paper was first invented, human beings have found ways of using it to pass on messages to each other. Books, magazines and newspapers are now part of our everyday lives, but with the invention of the Internet and the speed of new technological advances, the world of publishing is changing. Online newspapers and magazines, blogs, and even e-book readers are changing the way we get information. But will we ever stop picking up a good old-fashioned newspaper? E-publishing versus paper publishing – who will win?

B Work in teams. Team A prepares a list of the advantages of traditional publishing over e-publishing. Team B prepares a list of the advantages of e-publishing over traditional publishing. Use your dictionary, the Internet and your teacher to help you.

C Debate your ideas. Which team has the most convincing position?

An e-book, the electronic equivalent of a printed book

Unit 22 Multimedia

1 Multiple forms of media

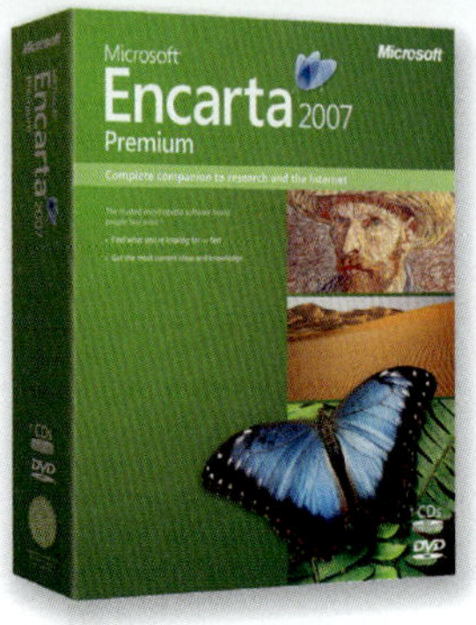

In pairs, discuss these questions.

1 What different types of content are combined in multimedia applications?

2 How many products can you think of that incorporate multimedia? Make a list.

2 Components and system requirements

A Listen to a sales assistant in a computer shop explaining to a customer the system requirements needed to run multimedia software. Which answers (a or b) best describe what she says?

1 Multimedia is defined as
  - **a** the integration of video and telecommunications with traditional computing.
  - **b** the integration of text, graphics, audio, video and animation in a single application.

2 With multimedia encyclopedias,
  - **a** you have more fun but you learn more slowly.
  - **b** you get much more involved than with print encyclopedias.

3 Interactive games
  - **a** use multimedia and virtual reality features.
  - **b** do not require much RAM memory.

B Listen again and complete this diagram of a multimedia system.

Multimedia system

Hardware

- (1) .... CPU and a lot of RAM
- **Peripherals**
  - (2) .... monitor
  - Large hard drive
  - DVD drive
  - (3) .... card
  - Stereo (4) .... or (5) ....
  - (6) ....

Software

- Windows Media Player or Apple QuickTime
- Adobe Flash for (7) .... media
- Audio, video and (8) .... software
- Hypermedia databases
- CD and DVD (9) ....

3 Multimedia magic!

A Read the text and match the headings (1–4) with the gaps at the start of each paragraph (a–d).

1 Sound, Music, MIDI
2 Products full of pictures, action and sound
3 Creating and editing movies
4 The potential of multimedia

Multimedia magic!

a ....

Multimedia applications are used in all sorts of fields. For example, museums, banks and estate agents often have information kiosks that use multimedia; companies produce training programs on optical discs; businesspeople use Microsoft PowerPoint to create slideshows; and teachers use multimedia to make video projects or to teach subjects like art and music. They have all found that moving images and sound can involve viewers emotionally as well as inform them, helping make their message more memorable.

The power of multimedia software resides in **hypertext, hypermedia** and **interactivity** (meaning the user is involved in the programme). If you click on a hypertext link, you can jump to another screen with more information about a particular subject. Hypermedia is similar, but also uses graphics, audio and video as hypertext elements.

b ....

As long as your computer has a **sound card**, you can use it to capture sounds in digital format and play them back. Sound cards offer two important capabilities: a built-in stereo synthesizer and a system called **MIDI**, or **M**usical **I**nstrument **D**igital **I**nterface, which allows electronic musical instruments to communicate with computers. A **Digital Audio Workstation** (**DAW**) lets you mix and record several tracks of digital audio.

MIDI allows your computer to communicate with electronic keyboards and other devices

You can also listen to music on your PC, or transfer it to a portable **MP3** player. MP3 is short for **MPEG audio layer 3**, a standard format that compresses audio files. If you want to create your own MP3 files from CDs, you must have a **CD ripper**, a program that extracts music tracks and saves them on disk as MP3s.

Audio is becoming a key element of the Web. Many radio stations broadcast live over the Internet using **streaming audio technology**, which lets you listen to audio in a continuous stream while it is being transmitted. The broadcast of an event over the Web, for example a concert, is called a **webcast**. Be aware that you won't be able to play audio and video on the Web unless you have a **plug-in** like RealPlayer or QuickTime.

c ....

Video is another important part of multimedia. **Video computing** refers to recording, manipulating and storing video in **digital format**. If you wanted to make a movie on your computer, first you would need to capture images with a **digital video camera** and then transfer them to your computer. Next, you would need a **video editing** program like iMovie to cut your favourite segments, re-sequence the clips and add transitions and other effects. Finally, you could save your movie on a DVD or post it on websites like YouTube and Google Video.

d ....

Multimedia is used to produce dictionaries and encyclopedias. They often come on DVDs, but some are also available on the Web. A good example is the Grolier Online Encyclopedia, which contains thousands of articles, animations, sounds, dynamic maps and hyperlinks. Similarly, the Encyclopedia Britannica is now available online, and a concise version is available for iPods, PDAs and mobile phones. Educational courses on history, science and foreign languages are also available on DVD. Finally, if you like entertainment, you'll love the latest multimedia video games with surround sound, music soundtracks, and even film extracts.

B Correct the technical mistakes in these sentences.

1 Multimedia training software is distributed on magnetic disks.
2 You need to have MIDI on your computer to hear speech and music.
3 A stereo synthesizer allows your computer to communicate with electronic musical instruments.
4 A CD ripper converts CDs to live streams.
5 The Encyclopedia Britannica is only available on DVD.

C Match the words (1–5) with the definitions (a–e).

1 hypertext	a the process of manipulating video images
2 hypermedia	b text with links which take you to other pages
3 streaming	c a technique for playing sound and video files while they're downloading
4 webcast	d a live event broadcast over the Internet
5 video editing	e a form of enriched multimedia which allows an interactive and dynamic linking of visual and audio elements

4 *Language work: conditional sentences*

A Look at the HELP box and then complete these sentences with the correct form of the verbs in brackets.

1 If you (bring) .... your digital video camera, we can make a movie on my PC.
2 You won't be able to play those video files if you (not have) .... the correct plug-in.
3 If the marketing manager (have) .... PowerPoint, she could make more effective presentations.
4 If I could afford it, I (buy) .... a new game console.
5 If I had the money, I (invest) .... in some new multimedia software.

HELP box

Conditional sentences

We use conditional sentences to express that the action in the main clause can only take place if a certain condition is fulfilled (see below for examples). They are introduced by **if, unless** and **as long as**. **Unless** means *if not* and **as long as** means *provided/providing (that)*.

*You won't be able to play audio and video on the Web **unless** you have a plug-in like RealPlayer or QuickTime.*
(= if you don't have a plug-in ...)

There are two types of conditional sentence.

- The first conditional (for real or possible situations).

  If A happens (present simple) — B will happen (**will** in positive or negative + verb)

  *If you **like** entertainment, you **will love** the latest multimedia video games.*

  In the main clause, we can also have a modal (for example, **can** or **must**) or an imperative.

  *If you **want** to create your own MP3 files from your CDs, you **must have** a CD ripper.*

- The second conditional (for more hypothetical situations).

  If A happened (past simple) — B would happen (**would** in positive or negative + verb)

  *If you **wanted** to make a movie on your computer, first you **would** / **you'd** need to ...*

  In the main clause, we can also use other modals (e.g. **could, should, might**), depending on the meaning.

  If the verb **be** appears in the **if** clause, we often use **were** instead of **was**, even if the pronoun is **I**, **he**, **she** or **it**.

  *If I **were** you, I'd get a new MP3 player.*

B **In pairs, discuss these questions. Use the second conditional.**

What would you do if …

1 you had a digital video camera?
If I had a digital video camera, I'd …
2 you had a home recording studio?
3 you couldn't afford an iPod but you wanted an MP3 player?
4 you won the lottery?
5 someone stole your laptop?

5 *Applications of multimedia*

A **Match the descriptions (1–5) with the pictures (a–e).**

1 Virtual reality
2 Distance learning
3 A business presentation
4 A touch screen information kiosk
5 An MMS mobile phone

a ....

b ....

c ....

d ....

e ....

B **In pairs, discuss how multimedia is used in the situations above and then present your ideas to the rest of the class. Look at the *Useful language* box to help you.**

Useful language

In distance learning, multimedia is used to …
Information kiosks take advantage of multimedia in order to …
In virtual reality, the use of multimedia allows you to …
With 3G mobile phones, you can …
Slide presentations integrate a wide range of media, such as …

C **A friend who writes a blog has asked you to contribute a post about the use of multimedia now and in the future. Write a post (80–120 words) summarizing what multimedia is, what it can do, and your predictions for what it might be able to do in the future. Try to use at least two conditional sentences.**

Unit 23 Web design

1 *A typical home page*

In pairs, discuss these questions.

1 Why do companies have websites?
2 What is the difference between a *website* and a *web page*?
3 What is a home page?
4 Do you have a blog or personal website? Describe the home page to your partner.

The Yahoo! home page

2 *Web page design*

A Read the text on page 115 and find the following.

1 the language used to create web documents
2 the type of software that lets you design web pages without writing HTML codes
3 the format invented by Adobe to distribute text files over the Internet
4 a method of displaying multiple HTML documents in the same browser window
5 three common graphics formats used on websites
6 three popular formats used to store and play back video

Web page design

HTML and web editors

The code behind most web pages is **HTML** (**h**yper**t**ext **m**arkup **l**anguage), which consists of commands called **tags**. Tags are placed around pieces of text to tell the web browser how to display text or pictures. You can view the source HTML code of a web page by choosing the *Page Source* option in your web browser. But you needn't learn HTML in order to build your own website. Instead, you can use a word processor with web design capabilities or a dedicated **web editor** like Macromedia Dreamweaver or Microsoft FrontPage. Web editors are user-friendly and WYSIWYG (*What You See Is What You Get*). Different buttons and menu items let you design a page without writing HTML.

HTML files have this basic structure:

```
start with <HTML>
have heading text that begins with <HEAD>, which has a
title enclosed between tags and ends with </HEAD>
have body text that starts with <BODY>, where you place
the contents of the actual document (i.e. text, images,
links, etc.) and ends with </BODY>
end with </HTML>
You can create links to other web pages by using the tag <A
HREF="filename">active text</A>
```

Some basic HTML source code

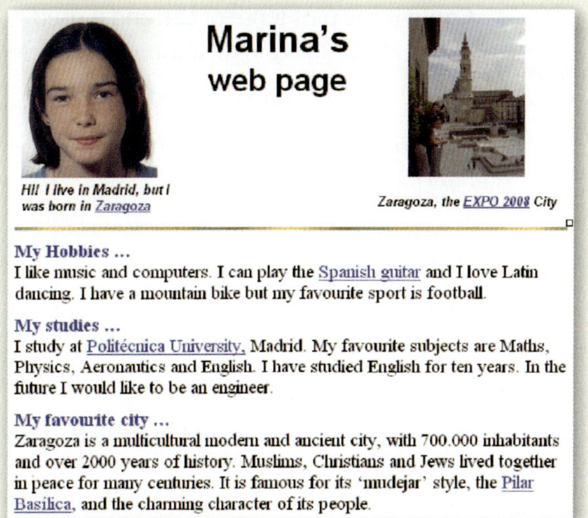

HTML file displayed as a web page

Web page elements

There are a number of different elements that you can use on a web page:

- **Text** – displayed in a variety of fonts and sizes. Most text files are available in two formats: HTML or **PDF** (the **p**ortable **d**ocument **f**ormat that can be viewed with Acrobat Reader).
- **Background** – the underlying colours and patterns of a web page
- **Tables** – with columns and rows, used to position images and text on a page
- **Frames** – rectangular areas that allow the display of different pages in the same browser window
- **Cascading Style Sheets (CSS)** – a mechanism for adding styles to web documents. You could use HTML code to specify the font, text styles and background colour. Nowadays, however, it is more common to use CSS. This makes it easy to apply presentation changes across a website.
- **Graphics, clip art, icons, background templates, wallpaper,** and **transparent images** – common formats are **.jpg** (joint photographic experts group), ideal for pictures with many colours, **.gif** (graphics interchange format), ideal for pictures with fewer colours, and **.png** (portable network graphics), which supports 16 million colours.
- **Hyperlinks** – highlighted text or pictures (buttons, image maps, etc.) that act as links to other pages. If you want to share information with people, you can use **RSS feeds** and provide readers with a link to the feed. RSS allows subscribers to receive updates of blogs, news, podcasts, etc. Before **going live**, you should check that all the links work.

Audio, video and animation

Many websites now incorporate audio files, and if you're designing a site, you may like to insert songs, podcasts, etc. The most common audio formats are: **.wav** (Windows wave audio format), **.ra** (RealAudio file) and **.mp3** (MPEG-1 Audio Layer-3).

Full-motion video is stored in these formats: **.avi** (audio video interleave), **.mov** (QuickTime movie) and **.mpg** (moving picture experts group).

If you want to inject something special into your web pages, you can use Adobe Flash to include **interactive animations** and **streaming audio**. Additionally, you can insert Java applets – small programs that enable the creation of interactive files. Animations are made up of a series of independent pictures put together in sequence to look like moving pictures. To see or hear all these files, you must have the right **plug-in**, an auxiliary program that expands the capabilities of your web browser.

B Read the text again and then match the sentence beginnings (1–6) with the correct endings (a–f).

1 Instructions in HTML
2 Cascading Style Sheets are the way
3 A hyperlink is any clickable text,
4 A plug-in is a small program
5 Java applets are used to provide
6 RSS feeds are summaries of web content

a image or button that takes you to another place on the Web.
b used for handling audio, video and animation files.
c are called *tags*.
d interactive features to web applications.
e to define the presentation of web pages, from fonts and colours to page layout.
f published in the Really Simple Syndication format for download.

3 *Language work: modal verbs*

A Underline all the modal verbs in the text on page 115 and then look at the HELP box. Which modal verb from the HELP box does not appear in the text? Can you think of any other modal verbs?

HELP box

Modal verbs

We use modal verbs to add extra meaning to the main verb. They are followed by infinitive without *to*. Modal verbs are used in the following ways:

- To express a possibility

  You ***can/could*** *use Adobe Flash to include interactive animations.*

  You ***may*** *like to insert songs, podcasts, etc.*

  The price of Dreamweaver ***might*** *go down next month.*

  Can and **could** are often interchangeable when talking about possibility. **May** and **might** are used to express weaker possibilities and often come before the verb **like** to mean *It is possible you will like.*

- To ask for permission

  Can/Could/May *I use your mobile phone?*

  May is more formal than **can** or **could**.

- To talk about ability

  They are looking for artists who ***can*** *draw and design web pages.*

  Could is the past tense of **can** and is used to talk about ability in the past.

- To talk about obligation or necessity

  To see or hear all these files, you ***must*** *have the right plug-in.*

  ... you ***needn't*** *learn HTML in order to build your own website.*

  Needn't means *don't need to* or *don't have to* and is used to express a lack of obligation.

- To give advice (see Unit 7)

  Before going live, you ***should*** *check that all the links work.*

B Complete these sentences with suitable modal verbs from the HELP box. There may be more than one possible answer.

1 With Java, I .... include some attractive banners on my website.
2 With a web editor, you .... create a web document easily.
3 These days, you .... learn how to use complicated HTML codes. Modern web design software is user-friendly and converts a visual layout into HTML code.
4 Once live, you .... update your website regularly.
5 To view a PDF file, you .... have Adobe Acrobat Reader.
6 Websites with graphics are more inviting than those written in plain text, so you .... like to insert some graphics into your documents.
7 .... I use your laptop? I need to print out this report.

C In pairs, discuss at least two things

1 you can now do more easily because of the Internet.
2 you could do better if you had a faster internet connection.
3 that may/might happen to the Internet in the next ten years.
4 you must consider when designing a website.
5 you should take into account when choosing which PC to buy.

4 *Designing a website*

A In pairs, think about your favourite websites and discuss these questions.

1 Do you like the way they are designed? Give reasons for your answer.
2 What elements do you think a good website should have? Make a list.

B Listen to an interview with a web designer describing how to design a website and put these steps into the correct order.

- ☐ Write and format the text
- ☐ 1 Decide the content and structure for the website
- ☐ Publish the website
- ☐ Insert computer graphics and sounds
- ☐ Keep the website updated
- ☐ Link related pages to each other using hyperlinks

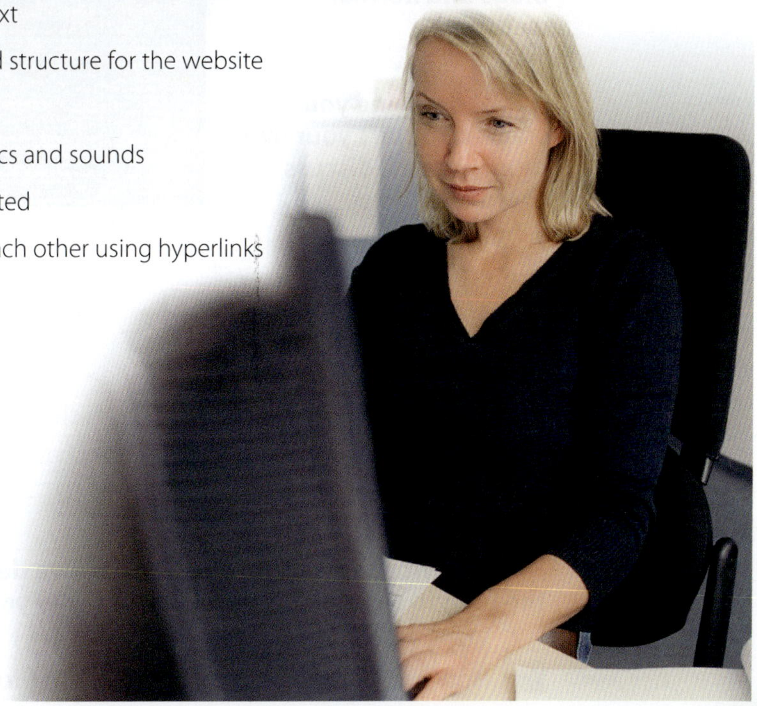

A web designer at work

C **Listen again and decide whether these design guidelines are right or wrong. Tick the correct box.**

		Right	Wrong
1	Plan your website carefully.	☐	☐
2	Use a web editor. It will make it easier to create your pages.	☐	☐
3	Insert photos or animations just to make the pages look attractive.	☐	☐
4	Place a large number of graphics on your pages.	☐	☐
5	Use very bright colours.	☐	☐
6	Put a lot of links on one page.	☐	☐
7	Check that all the links on your web pages are correct.	☐	☐
8	Once they are published, update your pages regularly.	☐	☐

D **In small groups, collect information about your college or company and design a home page for it. Follow the instructions from the interview with the web designer.**

5 *Blogs*

A **In pairs, discuss these questions.**

1 What is a blog?

2 Which blogs do you read regularly?

B **Look at the screenshot from tpsreport.co.uk, a popular gaming blog. Can you see any design differences between blogs and normal websites?**

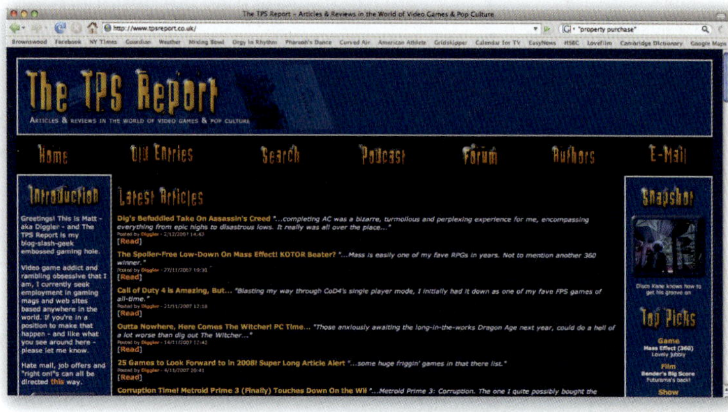

A screenshot from www.tpsreport.co.uk

C **Imagine you wanted to start your own blog. In pairs, discuss these questions.**

1 Why would you start your own blog – to write a diary of your thoughts or to share your expertise on a particular topic?

2 What types of media would you include – text, photos, video, audio (including podcasts)?

3 Would you insert links to other blogs? Which ones?

4 Would you focus on a particular subject or have a mix of several topics?

5 Which site would you use to host your blog?

D **Write an entry for the blog you've described in C (80–100 words). Introduce the blog to the world and talk about why you've started it.**

Now visit www.cambridge.org/elt/ict for an online task.

7 Programming / Jobs in ICT

Learning objectives

In this module, you will:

- study basic concepts in programming.
- learn and use vocabulary connected with programming and become familiar with word families.
- ask and answer questions about computer languages.
- learn and use the basic vocabulary associated with the Java language.
- talk about your personal experience of using computers.
- practise the use and pronounciation of the *-ed* form of verbs.
- discuss the personal qualities and professional skills needed for a job in ICT.
- learn how to understand job advertisements.
- learn how to write a CV and a letter applying for a job.

Unit 24 Program design and computer languages

1 Programming

A In pairs, discuss what you think *programming* is.

B Look at the definition of *programming* in the Glossary. Is it similar to yours?

```
#include <stdio.h>
main( )
{
printf("good morning\n");
}
```

This C program tells the computer to print the message 'good morning'

2 Steps in programming

A Match the words (1–5) with the definitions (a–e).

1 flowchart
2 source code
3 compiler
4 machine code
5 debugging

a Program instructions written in a particular computer language
b The techniques of detecting and correcting errors (or bugs) which may occur in programs
c A diagram representing the successive logical steps of the program
d A special program which converts the source program into machine code – the only language understood by the processor
e The basic instructions understood by computers; it consists of 1s and 0s (binary code)

B Listen to Andrea Finch, a software developer, talking to a group of students on a training course about how a program is written and check your answers to A.

C Listen again and put these steps into the correct order.

- [ ] Write instructions in a programming language
- [ ] Prepare documentation
- [1] Understand the problem and plan a solution
- [ ] Make a flowchart of the program
- [ ] Compile the program (to turn it into machine code)
- [ ] Test and debug the program

D Listen again and make detailed notes. In pairs, use your notes to write a short explanation of what each step in C means.

3 *Computer languages*

A Read the text. How many high-level computer languages are mentioned?

Computer languages

Unfortunately for us, computers can't understand spoken English or any other natural language. The only language they can understand directly is **machine code**, which consists of 1s and 0s (binary code).

Machine code is too difficult to write. For this reason, we use symbolic languages to communicate instructions to the computer. For example, **assembly languages** use abbreviations such as ADD, SUB, MPY to represent instructions. The program is then translated into machine code by a piece of software called an **assembler**. Machine code and assembly languages are called **low-level languages** because they are closer to the hardware. They are quite complex and restricted to particular machines. To make the programs easier to write, and to overcome the problem of intercommunication between different types of computer, software developers designed **high-level languages**, which are closer to the English language. Here are some examples:

- **FORTRAN** was developed by IBM in 1954 and is still used for scientific and engineering applications.
- **COBOL** (**Co**mmon **B**usiness **O**riented **L**anguage) was developed in 1959 and is mainly used for business applications.
- **BASIC** was developed in the 1960s and was widely used in microcomputer programming because it was easy to learn. **Visual BASIC** is a modern version of the old BASIC language, used to build graphical elements such as buttons and windows in Windows programs.
- **PASCAL** was created in 1971. It is used in universities to teach the fundamentals of programming.
- **C** was developed in the 1980s at AT&T. It is used to write system software, graphics and commercial applications. **C++** is a version of C which incorporates object-oriented programming: the programmer concentrates on particular things (a piece of text, a graphic or a table, etc.) and gives each object functions which can be altered without changing the entire program. For example, to add a new graphics format, the programmer needs to rework just the graphics object. This makes programs easier to modify.
- **Java** was designed by Sun in 1995 to run on the Web. Java applets provide animation and interactive features on web pages. (See Unit 25)

Programs written in high-level languages must be translated into machine code by a **compiler** or an **interpreter**. A compiler translates the source code into **object code** – that is, it converts the entire program into machine code in one go. On the other hand, an interpreter translates the source code line by line as the program is running.

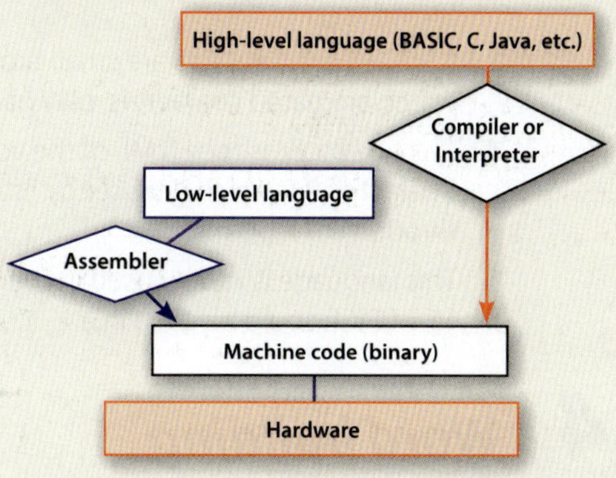

It is important not to confuse **programming languages** with **markup languages**, used to create web documents. Markup languages use instructions, known as **markup tags**, to format and link text files. Some examples include:

- **HTML**, which allows us to describe how information will be displayed on web pages.
- **XML**, which stands for **EX**tensible **M**arkup **L**anguage. While HTML uses pre-defined tags, XML enables us to define our own tags; it is not limited by a fixed set of tags.
- **VoiceXML**, which makes Web content accessible via voice and phone. VoiceXML is used to create voice applications that run on the phone, whereas HTML is used to create visual applications (for example, web pages).

```
<xml>
< name> Andrea Finch </name>
< homework> Write a paragraph describing
the C language </homework>
</xml>
```

In this XML example we have created two new tags: <name> and <homework>

B Read the text again and answer these questions.

1 Do computers understand human languages? Why? / Why not?
2 What is the function of an *assembler*?
3 Why did software developers design high-level languages?
4 Which language is used to teach programming techniques?
5 What is the difference between a *compiler* and an *interpreter*?
6 Why are HTML and VoiceXML called *markup* languages?

C Complete these sentences with a computer language from the text.

1 .... allows us to create our own *tags* to describe our data better. We aren't constrained by a pre-defined set of tags the way we are with HTML.
2 IBM developed .... in the 1950s. It was the first high-level language in data processing.
3 .... applets are small programs that run automatically on web pages and let you watch animated characters, play games, etc.
4 .... is the HTML of the voice web. Instead of using a web browser and a keyboard, you interact with a voice browser by listening to pre-recorded audio output and sending audio input through a telephone.
5 This language is widely used in the business community. For example, the statement ADD VAT to NET-PRICE could be used in a .... program.

4 *Word building*

Look at the words in the boxes. Are they nouns, verbs or adjectives? Write *n*, *v* or *adj* next to each word. There may be more than one possible answer. Complete the sentences with words from the boxes.

program	programmers	programming	programmable ....

1 .... is the process of writing a program using a computer language.
2 A computer .... is a set of instructions that tells the computer how to do a specific task.
3 Most computer .... make a plan of the program before they write it.
4 A .... keyboard allows the user to configure the layout and meaning of the keys.

compile	compiler	compilation ....

5 Programs written in a high-level language require .... – that is, translation into machine code, the language understood by the processor.
6 A source program is converted into machine code by software called a ....
7 Programmers usually .... their programs to generate an object program and diagnose possible errors.

bug	debug	debugger	debugging ....

8 Any error or malfunction of a computer program is known as a ....
9 A .... is a program used to test and .... other programs.
10 The process of going through the code to identify the cause of errors and fixing them is called ....

5 Language work: the infinitive

A Look at the HELP box and then make sentences using these prompts.

1 not easy / write instructions in COBOL

It's not easy to write instructions in COBOL.

2 expensive / set up a data-processing area

3 advisable / test the programs under different conditions

4 unusual / write a program that works correctly the first time it's tested

5 important / use a good debugger to fix errors

6 easy / learn Visual BASIC

B Choose the correct words (a–c) to complete these sentences.

1 We use high-level languages because machine code is too difficult ...., understand and debug.

a read **b** reading **c** to read

2 I went on the course .... how to be a better programmer.

a learn **b** to learn **c** for to learn

3 I'm not interested in .... that computer language.

a learn **b** learning **c** to learn

4 He refuses .... the project with me.

a do **b** doing **c** to do

5 The engineers warned the employees not .... the cables.

a touch **b** touching **c** to touch

6 They may not .... to the conference.

a come **b** coming **c** to come

7 Spyware can make your PC .... more slowly.

a perform **b** performing **c** to perform

8 This program is too slow .... the simulation.

a do **b** to do **c** for doing

HELP box

The infinitive

The infinitive with *to* is used in the following ways:

- To express purpose

  We use symbolic languages ***to communicate*** *instructions to the computer.*
  (= *in order to communicate* ...)

  Not: ... ~~for~~ *to communicate*

- After adjectives

  BASIC was widely used in the past because it was ***easy to learn****.*

  Machine code is too ***difficult to write****.*
  (= *not easy enough to write*)

- After certain verbs (e.g. **afford**, **demand**, **plan**, **agree**, **expect**, **promise**, **appear**, **hope**, **refuse**, **arrange**, **learn**, **try**, **decide**, **manage**)

  A lot of companies are now ***trying to develop*** *voice applications for web access.*

- After the object of certain verbs (e.g. **advise**, **encourage**, **allow**, **expect**, **tell**, **ask**, **invite**, **want**, **enable**, **order**, **warn**)

  HTML ***allows us to describe*** *how information will be displayed on web pages.*

The bare infinitive (without *to*) is used in the following ways:

- After modal verbs (e.g. **can**, **could**, **may**, **might**, **will**, **would**, **must**, **should**)

  Unfortunately, computers ***can't understand*** *spoken English.*

  High-level languages ***must be*** *translated into machine code.*

- After the object with the verbs **make** and **let**

  Programs ***make computers perform*** *specific tasks.*

C **In pairs, discuss something**

1 you can't afford to buy at the moment.
2 you've arranged to do this weekend.
3 you've learnt to do in the last year.
4 you'd advise someone to do before buying a new PC.
5 you'd expect to be included with an anti-virus package.
6 you can do with Java applets.

6 *Visual BASIC and VoiceXML*

A Work in pairs. Student A reads about Visual BASIC, Student B reads about VoiceXML. Try not to look at your partner's text. Complete your part of the table.

Student A

Visual BASIC was developed by Microsoft in 1990. The name **BASIC** stands for Beginner's All-purpose Symbolic Instruction Code. The adjective **Visual** refers to the technique used to create a graphical user interface. Instead of writing a lot of instructions to describe interface elements, you just add pre-defined objects such as buttons, icons and dialog boxes. It enables programmers to create a variety of Windows applications.

Student B

VoiceXML (EXtensible Markup Language) was created in 2000 to make web content accessible via the telephone. For input, it uses voice recognition. For output, it uses pre-recorded audio content and text-to-speech. Applications:

- voice portals, where you can hear information about sports, news, traffic, etc.
- voice-enabled intranets (private networks)
- voice e-commerce
- home appliances controlled by voice

	Visual BASIC	VoiceXML
What does Visual BASIC / VoiceXML stand for?		
When was it developed?		
What are its main features?		
What is it used for?		

B **Ask your partner about the other language and complete the table.**

1 Java applets

A Match the examples of Java programs, known as *applets*, (a–e) with the descriptions (1–5).

1 This Land Rover applet allows you to change the look of the vehicle.

2 The Pythagoras theorem applet gives the proof of the Pythagorean theorem without words. It allows you to manipulate triangles and go through the steps of the geometrical proof.

3 The Jman for Java applet permits medical researchers to view sequential MRI (**M**agnetic **R**esonance **I**mages) of the brain.

4 An analogue clock applet displays the time according to the web user's computer and lets you set the colours and style of the hands and numbers.

5 A banner applet displays graphic images on websites in order to advertise products or services.

a ....

b ....

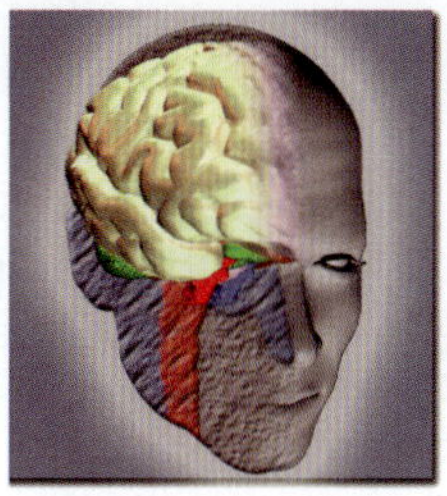

c ....

d ....

e ....

B Match the terms (1–5) with the definitions (a–e).

1 Java

2 applet

3 plug-in

4 platform-independent

5 object-oriented programming

a an auxiliary program that enables web browsers to support new content, for example animation

b software that can run on any operating system

c an island in Indonesia, coffee (in American slang), and a programming language for internet applications

d a computer programming technique that allows the creation of objects that interact with each other and can be used as the foundation of others; used to create graphical user interfaces

e a small Java application, usually designed to run automatically within a web page

The Java logo

2 *The Java language*

A These statements about Java are all false. Read the text and correct them.

1 Java was invented by Microsoft.
2 With the interpreter, a program is first converted into Java bytecodes.
3 Java is not compatible with most computing platforms.
4 The Java language is single-threaded, one part executing at a time.
5 Java has no competitors.
6 Flash files are called *animations*.

The Java language

Java is a programming language developed by Sun Microsystems, specially designed to run on the Web. Java programs (called **applets**) let you watch animated characters and moving text, play music, and interact with information on the screen (for example, control animations and select options).

Characteristics of the Java language

Java is an **object-oriented** language, similar to C++, but more dynamic and simplified to eliminate possible programming errors. A Java program is both compiled and interpreted (see Unit 24). First, the source code (a file with a **.java** extension) is compiled and converted into a format called bytecode (a file with a **.class** extension), which can then be executed by a Java interpreter (see Fig. 1). Compiled Java code can run on most computers because there are Java interpreters, known as **Java Virtual Machines**, for most operating systems.

Java is **multi-threaded**, meaning a Java program can have multiple threads (parts) – that is, many different things processing independently and continuously. This enables the program to make the best use of available CPU power.

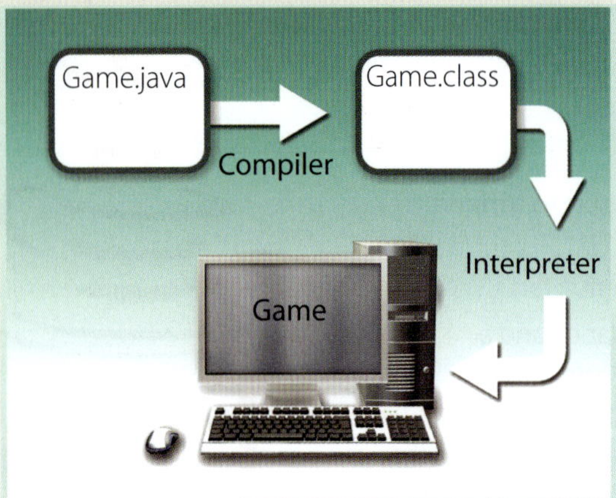

Fig. 1

Why is Java popular?

Most programmers like Java because it allows them to write applets which make web pages more interactive and attractive. They can create graphical objects (for example, bar charts and diagrams) and new controls (for example, check boxes and push buttons with special properties). A web page that uses Java can have sounds that play in real time, music that plays in the background, cartoon-style animations, real-time video and interactive games.

The Java Micro Edition platform (**Java ME**) is used in mobile devices. It provides flexible tools to create applications that run on mobile phones, PDAs, TV set-top boxes and printers. Nowadays, most phones are configured to use Java games.

Alternatives to Java

One alternative to Java is Microsoft's **C#**, pronounced 'C sharp', a **.NET** language based on C++ with elements from Visual Basic and Java. There are no substantial differences between C# and Java. When software developers do measurements on pieces of code, sometimes Java is faster, sometimes C# is.

Another competitor is Adobe **Flash** technology, which supports graphics, a scripting language called ActionScript, and the streaming of audio and video. Flash is used to create animation and advertisements, to integrate video into web pages, and to develop rich internet applications such as portals. **Flash files**, traditionally called **flash movies**, have a **.swf** file extension. They may be an object on a web page or be played in the stand-alone Flash Player.

B Match the words (1–6) with the words (a–f) to make technical terms from the text.

1 Java	4 web	a applet	d system
2 operating	5 source	b page	e object
3 programming	6 graphical	c code	f language

C Complete the sentences with words from the box.

interpreted	animated	configured	used	pronounced	object-oriented	compiled

1 Java lets you watch .... characters on web pages.
2 Java is an .... language, similar to C++ but more dynamic.
3 First, the source code of a Java program is .... into an intermediate format called *bytecode*. This is then .... by any system possessing a Java interpreter.
4 The Java ME platform is widely .... in mobile devices.
5 Nowadays, most mobile phones are .... to use Java games.
6 Microsoft's C# is a simplified version of C and C++ for the Web. It's .... 'C sharp'.

3 *Language work: the -ed form*

A Look at the HELP box and then put these verbs into the correct column.

stopped	asked
described	decided
produced	called
watched	executed
published	object-oriented
programmed	persuaded
configured	converted
arranged	designed

/t/	/d/	/ɪd/

HELP box

The *-ed* form

We use the **-ed** form in the following ways:

- To make the past simple (affirmative) of regular verbs

  *Sun Microsystems **developed** Java in 1995.*

  Remember that not all verbs in the past simple end in **-ed**. See page 166 for a list of irregular verbs. See Unit 19 for more about the past simple.

- To make the past participle of regular verbs

  *Flash is **used** to create animation.*

- To make the adjectival form of some verbs

  *Java applets let you watch **animated** characters.*

The **-ed** is pronounced as:

- /t/ after voiceless sounds: /p/, /k/, /θ/, /s/, /f/, /ʃ/or /tʃ/. (e.g. *developed, talked, pronounced*)
- /d/ after voiced sounds: /b/, /g/, /ð/, /z/, /v/, /dʒ/, /l/, /r/; nasal consonants: /m/, /n/, /ŋ/; and vowels (e.g. *compiled, designed, simplified*)
- /ɪd/ after /t/ or /d/ (e.g. *interpreted, multi-threaded*)

B **Complete this extract from a lecture handout about Java with the correct form of the verbs in the box.**

call	be	begin	can	decide	rename	have	support	develop	base

The idea for Java started in 1990, when a team of software engineers at Sun Microsystems (1) .... to create a language for a handheld device that could control and interact with various kinds of electronic appliances, ranging from Nintendo Game Boys to VCRs and TV set-top boxes. They (2) .... an object-oriented programming language that one of the engineers, James Gosling, (3) .... *Oak*, after the tree outside his window. The device even (4) .... an animated character named *Duke*, who would go on to become Java's mascot.

With the advent of the Web in 1993, the company made a web browser (5) .... on the Oak language. Later on, this language was adapted to the Internet and (6) .... *Java*. The 1.0 version of Java was officially introduced by Sun in May 1995.

At that time, web pages (7) .... only display text, pictures and hyperlinks. With the arrival of Java, web designers (8) .... able to include animation and interactive programs on web pages. The first major application created with Java was the HotJava browser. The Java language (9) .... to attract serious attention from the internet community and was soon (10) .... by Netscape Navigator and MS Internet Explorer. Today, Java is a hot technology that runs on multiple platforms, including smart cards, embedded devices, mobile phones and computers.

C **Listen to an extract from the lecture and check your answers to C. Listen carefully to the pronunciation of the verbs that end in *-ed*.**

4 *Your experience with computers*

A **Make notes about the different stages in your computer history. Add more stages if you want to.**

Example: *1990: Played my first computer game. It was ...*

Possible stages:

- First computer game
- First computer lesson at school/college
- First programming language learnt
- First software used
- First computer course/qualification
- First job involving computers
- First steps on the Internet
- First chat online

B **Ask a partner about their computer history. Look at the *Useful language* box to help you.**

Useful language

When did you first ...?

How long ago did you ...?

How old were you when ...?

I started ... in...

I learnt ... when I was ...

I didn't use the Internet until ...

Unit 26 Jobs in ICT

1 IT professionals

A Complete these definitions with jobs from the box.

software engineer	computer security specialist	blog administrator	help desk technician
DTP operator	hardware engineer	network administrator	webmaster

1 A .... designs and develops IT devices.
2 A .... writes computer programs.
3 A .... edits and deletes posts made by contributors to a blog.
4 A .... uses page layout software to prepare electronic files for publication.
5 A .... manages the hardware and software that comprise a network.
6 A .... designs and maintains websites.
7 A .... works with companies to build secure computer systems.
8 A .... helps end-users with their computer problems in person, by email or over the phone.

B Listen to four people on a training course introducing themselves and talking about their jobs. Which job in A does each person do?

Speaker 1 .... Speaker 3 ....

Speaker 2 .... Speaker 4 ....

2 Job advertisements

A In pairs, read the two job advertisements on page 130 and tick (✓) the most important qualities and abilities (1–10) for each job. Add more to the list if you can. Which three things do you think are most important for each job?

		Senior programmer	DTP operator
1	logical reasoning	☐	☐
2	patience and tenacity	☐	☐
3	being good with figures	☐	☐
4	imagination	☐	☐
5	self-discipline	☐	☐
6	accuracy	☐	☐
7	leadership skills	☐	☐
8	efficiency	☐	☐
9	creativity	☐	☐
10	drawing skills	☐	☐

B Discuss if you would like to apply for one of the jobs. Give reasons for your answers.

DIGITUM-UK

SENIOR PROGRAMMER required by DIGITUM-UK, a leading supplier of business systems to the insurance industry.

You will be able to work on the full range of software development activities – analysis, design, coding, testing, debugging and implementation. At least two years' experience of COBOL or C++ is necessary.

As we are active in Europe, fluency in French, Italian or another European language is desirable.

Don't miss this opportunity to learn new skills and develop your career.

Send your CV to CHRIS SCOTT, PERSONNEL MANAGER, DIGITUM-UK, 75 PARKSHILL STREET, LONDON SW14 3DE

You can visit our website at www.digitum-uk.com

DTP operator

required for a leading financial magazine.

We are looking for a bright, competent QuarkXPress operator with at least three years' experience in design and layout. Skills in Photoshop, Freehand or Illustrator an advantage.

Ability to work in a team and to tight deadlines is vital.

Please apply in writing, with CV and samples of your work, to Tom Parker, Production Manager, Financial Monthly, Stockton Street, London EC1A 4WW
Or apply online:

Apply now

C Look at the online profile for Charles Graham. Which of the jobs above is most appropriate for him?

Charles Graham 22 years old

Professional summary

I graduated in 2004 with A levels in English, Art and Maths, and went on to do a course in graphic design and page layout at Highland Art School. Since 2006 I've been a graphic designer for PromoPrint, a company specializing in publishing catalogues and promotional material, and have used Adobe InDesign and other DTP software.

3 *A letter of application*

A Read the letter of application on page 131 and answer these questions.

1 Which job is Sarah Brown applying for?
2 Where did she see the advertisement?
3 How long has she been working as a software engineer?
4 What type of programs has she written?
5 When did she spend three months in Spain?

Dear Mr Scott,

I am writing to apply for the position of Senior Programmer, which was advertised on 28th March in *The Times*.

I graduated in May 2002 and did a work placement with British Gas as part of my degree. Before taking my present job I worked for a year with NCR. I stayed in this job (1) .... March 2004.

(2) .... the last three years I have been working as a software engineer for Intelligent Software. I have designed four programs in COBOL for commercial use, and (3) .... January I have been writing programs in C for use in large retail chains. These have been very successful and we have won several new contracts in the UK and Europe on the strength of my team's success.

Two years (4) .... I spent three months in Spain testing our programs and also made several visits to Italy, so I have a basic knowledge of Spanish and Italian. I now feel ready for more responsibility and more challenging work, and would welcome the opportunity to learn about a new industry.

I enclose my curriculum vitae. I will be available for an interview at any time.

I look forward to hearing from you.

Yours sincerely,

Sarah Brown

Sarah Brown

HELP box

for*, *since*, *ago*, *until

- We use **for** to refer to a period of time.
  I've lived in Liverpool ***for*** *five years.*
- We use **since** to refer to a point in time.
  I've been unemployed ***since*** *May 2005.*
- We use **ago** with the past simple to say when something happened. We put **ago** after the time period.
  I got married five years ***ago****.*
- We use **until** to mean *up to a certain time.*
  I stayed at high school ***until*** *I was 18.*

B Look at the HELP box and then complete the letter with *for, since, ago* or *until.*

4 *A job interview*

Chris Scott, the Personnel Manager at Digitum-UK, is interviewing Sarah Brown. Listen to part of the interview and complete his notes.

Name: Sarah Brown

Qualifications:

Degree in (1) ....

(Aston University)

Languages: Basic Spanish and Italian

Work experience:

NCR: (2) .... (one year)

Software for:

(3) ....

Programs for:

(4) ....

Database knowledge:

(5) ....

Present job: Works for Intelligent Software writing programs in COBOL and C.

Reasons for applying:

(6) ....

....

5 Language work: the present perfect

A Look at the HELP box and then choose the correct words in brackets to complete these sentences.

1 He ('s never liked / 's never been liking) Maths.
2 They ('ve worked / 've been working) on the project all day.
3 John ('s used / 's been using) the computer for hours – he looks really tired.
4 How many emails (have you written / have you been writing) today?
5 She ('s written / 's been writing) this essay since 9 o'clock.
6 They ('ve interviewed / 've been interviewing) five candidates today.

HELP box

Present perfect simple

We form the present perfect simple with **have/has** + past participle.

I've used *Microsoft Access for many years.*
I haven't used *Microsoft Access for years.*

We use this tense to talk about:

- States that started in the past and continue to the present.

  Since 2006, ***I've been*** *a computer operator for PromoPrint.*

- Past actions that continue to the present, where we put an emphasis on quantity (*how many*).
  I have designed *four programs in COBOL.*

- Personal experiences, especially with **ever** and **never**.

  Have you ever worked *with databases?*
  I've never worked *with databases.*

Present perfect continuous

We form the present perfect continuous with **have/has been** + present participle.

Since January ***I've been writing*** *programs in C.*

We use this tense to talk about:

- Actions which started in the past and are still happening.

  For the last three years ***I've been working*** *as a software engineer for Intelligent Software.*

- Past actions that continue to the present, where we put an emphasis on duration (*how long*).

  She's been working *all morning.*

Contrast with the past simple

We use the past simple to talk about events that happened at a specific time in the past that are now finished.

I ***graduated*** *in May 2003.*
Not: ~~***I have graduated***~~ *in …*
I ***stayed*** *in this job until March 2004.*
Two years ago, I ***spent*** *three months in Spain.*

B Put the verbs in brackets into the present perfect simple or past simple.

1 She (be) .... a software engineer since 2004.
2 After graduation I (work) .... for a year with NCR.
3 (you ever work) .... as an IT consultant?
4 I (lose) .... my PDA.
5 I (send) .... my CV last Monday. Have you received it yet?

C Make questions using these prompts. In pairs, ask and answer the questions.

1 ever / live or work in another country?
2 ever / have a bad job interview?
3 ever / do a job you hated?
4 how long / study English?
5 how long / use computers?
6 how many emails / receive today?
7 how many jobs / apply for this year?

6 *Applying for a job*

A Look at the job advertisement for a webmaster at eJupiter. María Quintana is interested in applying. Use her curriculum vitae on page 155 to write a letter of application. Follow these steps:

Paragraph one: reason for writing
I am writing to apply for the position of …

Paragraph two: education and training
I graduated in (date) …
I completed a course in …

Paragraph three: work experience
For the past X years I have been …
Since X I have been …

Paragraph four: personal skills
I spent X months in (country) … , so I have knowledge of (foreign languages).
I can …

Paragraph five: reasons why you are applying for this job
I now feel ready to … and would welcome the opportunity to …

Paragraph six: closing / availability for interview
I enclose … I look forward to … I will be available for an interview …

> **Vacancies at eJupiter.co.uk**
>
> **Webmaster**
>
> We are seeking a Webmaster for eJupiter.co.uk, a company dedicated to e-commerce.
>
> The successful candidate will manage our website. You will be responsible for making sure the web server runs properly, monitoring the traffic through the site, and designing and updating our web pages.
>
> Experience of using HTML and Java is essential. Experience of Adobe PDF and Photoshop is an advantage. The successful candidate will also have knowledge of web editors – MS FrontPage or equivalent.
>
> *Send your CV and a covering letter to James Taylor, eJupiter Computers, 37 Oak Street, London SW10 6XY*

B Write your own CV in English, using María's CV as a guide.

C Think of your ideal job and write a letter of application for it. If you prefer, look on the Internet for real jobs and practise applying for those.

Now visit www.cambridge.org/elt/ict for an online task.

8 Computers tomorrow

Learning objectives

In this module, you will:

- learn about different ICT systems.
- study the basics of networking.
- describe networks.
- learn and use phrasal verbs common in ICT.
- describe different game platforms and genres.
- give opinions about video games.
- learn and use adverbs.
- learn how to write a *For and Against* essay.
- make predictions about future trends.
- learn and use future forms.

Unit 27 Communication systems

1 *Information and communications technologies (ICT)*

A In pairs, discuss these questions.

1 What is an ICT system?

2 How many types of ICT system can you think of? Make a list.

3 How can a PC be connected to another computer?

B Label the pictures (1–7) with the ICT systems and services in the box.

Fax	GPS	Call centre	Digital radio	Teletext	Wearable computer	Digital TV

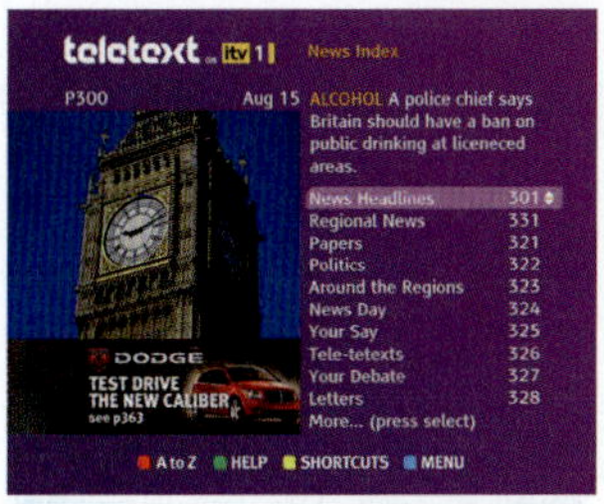

1 ....

2 ....

3 ....

4 ....

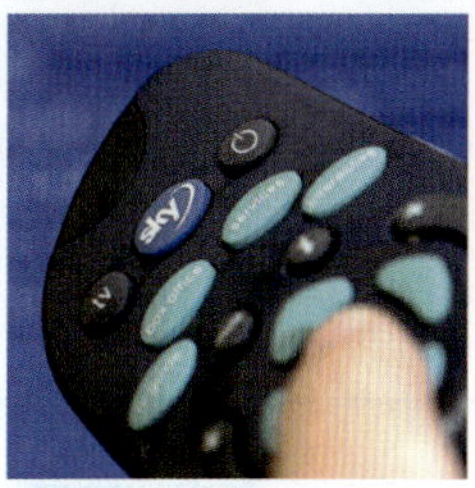

5 ....

6 ....

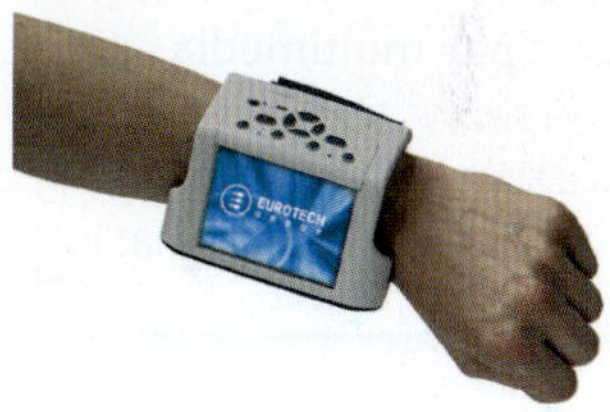

7 ....

C Complete these sentences with words and phrases from B and then read the text on page 136 to check your answers.

1 Digital Audio Broadcasting, or DAB, is the technology behind .... DAB is intended to replace FM in the near future.

2 .... are designed to be worn on the body or integrated into the user's clothing.

3 Most existing TV sets can be upgraded to .... by connecting a digital decoder.

4 My grandfather is 75 and he still watches .... on TV to find out share prices, weather forecasts and sports results.

5 I work in a .... I receive incoming calls with information inquiries. I also make outgoing calls for telemarketing.

6 Please complete this form and send it by .... or normal mail.

7 I have a .... navigation system in my car but I don't use it very often. My town is small and I know it well.

Channels of communication

What are telecommunications?

Telecommunications refers to the transmission of signals over a distance for the purpose of communication. Information is transmitted by devices such as the telephone, radio, television, satellite, or computer networks. Examples could be two people speaking on their **mobile phone**, a sales department sending a **fax** to a client, or even someone reading the **teletext** pages on TV. But in the modern world, telecommunications mainly means transferring information across the **Internet**, via modem, phone lines or wireless networks.

Because of telecommunications, people can now work at home and communicate with their office by computer and telephone. This is called **teleworking**. It has been predicted that about one third of all work could eventually be performed outside the workplace. In **call centres**, assistance or support is given to customers using the telephone, email or online chats. They are also used for **telemarketing**, the process of selling goods and services over the phone.

Digital TV and radio

In recent years, TV and radio broadcasting has been revolutionized by developments in satellite and digital transmission. **Digital TV** is a way of transmitting pictures by means of digital signals, in contrast to the analogue signals used by traditional TV. Digital TV offers interactive services and **pay multimedia** – that is, it can transmit movies and shows to TV sets or PCs on a pay-per-view basis. It is also **widescreen**, meaning programmes are broadcast in a native 16:9 format instead of the old 4:3 format. Digital TV provides a better quality of picture and sound and allows broadcasters to deliver more channels.

Digital Terrestrial TV is received via a **set-top box**, a device that decodes the signal received through the aerial. New technologies are being devised to allow you to watch TV on your mobile. For example, **DMB** (**D**igital **M**ultimedia **B**roadcasting) and **DVB-H** (**D**igital **V**ideo **B**roadcast-**H**andheld) can send multimedia (radio, TV and data) to mobile devices.

Audio programs (music, news, sports, etc.) are also transmitted in a digital radio format called **DAB** (**D**igital **A**udio **B**roadcasting).

Mobile communications

Thanks to wireless connectivity, mobile phones and **BlackBerrys** now let you check your email, browse the Web and connect with home or company intranets, all without wires.

The use of **GPS** in cars and PDAs is widespread, so you can easily navigate in a foreign city or find the nearest petrol station. In the next few years, GPS chips will be incorporated into most mobile phones.

Another trend is **wearable computers**. Can you imagine wearing a PC on your belt and getting email on your sunglasses? Some devices are equipped with a wireless modem, a keypad and a small screen; others are activated by voice. The users of wearable technology are sometimes even called *cyborgs*! The term was invented by Manfred Clynes and Nathan Kline in 1960 to describe cybernetic organisms – beings that are part robot, part human.

D Read the text again and find the following.

1 the device that allows PCs to communicate over telephone lines
2 the practice of working at home and communicating with the office by phone and computer
3 the term that refers to the transmission of audio signals (radio) or audiovisual signals (television)
4 five advantages of digital TV over traditional analogue TV
5 two systems that let you receive multimedia on your mobile phone
6 the term that means *without wires*
7 devices that deliver email and phone services to users on the move
8 the meaning of the term *cyborg*

2 Language work: the passive

A Look at the HELP box. How do you make the passive in your language? How different is it to English?

HELP box

The passive

We form the passive with the verb **be** + the past participle of the main verb. When we mention the agent, we use **by**.

The passive is often used in technical writing to give an objective tone.

- Present simple passive
  *Information **is transmitted by** devices such as the telephone, radio, TV or ...*
- Present continuous passive
  *New technologies **are being devised** to allow you to watch TV on your mobile.*
- Past simple passive
  The term **cyborg** ***was invented by** M Clynes and N Kline in 1960.*
- Past continuous passive
  *My TV **was being repaired**, so I couldn't watch the match.*
- Present perfect passive
  *It **has been predicted** that about one third of all work could eventually be performed outside the workplace.*
- Past perfect passive
  *The system **had been infected** by a virus.*
- Future simple passive
  *In the next few years, GPS chips **will** also **be incorporated** into most mobile phones.*
- Modal verbs in the passive
  *It has been predicted that about one-third of all work **could** eventually **be performed** outside the workplace.*

B Read the article and underline all the examples of the passive. What tenses are they?

A HACKER has been sent to jail for fraudulent use of credit card numbers. Nicholas Cook, 26, was arrested by police officers near a bank cashpoint last month.

Eight months earlier, he had been caught copying hundreds of computer programs illegally. After an official inquiry, he was accused of software piracy and fined £5,000.

It is reported that in the last few years Cook has been sending malware (malicious software) to phone operators and attacking mobile phones to steal business and personal information. Cook has now been sentenced to three years in prison for stealing passwords and obtaining money by credit card fraud.

Government officials say that new anti-hacking legislation will be introduced in the EU next year.

C Complete these sentences with the passive form of the verbs in brackets.

1 Microprocessors (make) .... of silicon.
2 Call centres (use) .... to deal with telephone enquiries.
3 In recent years, most mobile phones (equip) .... with Bluetooth.
4 GPS (develop) .... in the 1970s as a military navigation system.
5 Sorry about the mess – the computers (replace) .... at the moment.
6 In the near future, the Internet (access) .... more frequently from PDAs and mobile phones than from desktop computers.
7 Networks (can connect) .... via satellite.
8 I had to use my laptop this morning while my PC (fix) ....

3 *VoIP technology*

A Listen to an interview with Sue Reid, a specialist in telecommunications. What is her prediction about the future of VoIP?

B Listen again and answer these questions.

1 What exactly is VoIP?
2 Does the recipient need any special equipment?
3 What is an ATA? What is its function?
4 What is the advantage of Wi-Fi phones over mobile phones?
5 Do you need to have a VoIP service provider?
6 What is *spit*?

A wireless VoIP phone

C Using the diagram, explain VoIP technology in your own words.

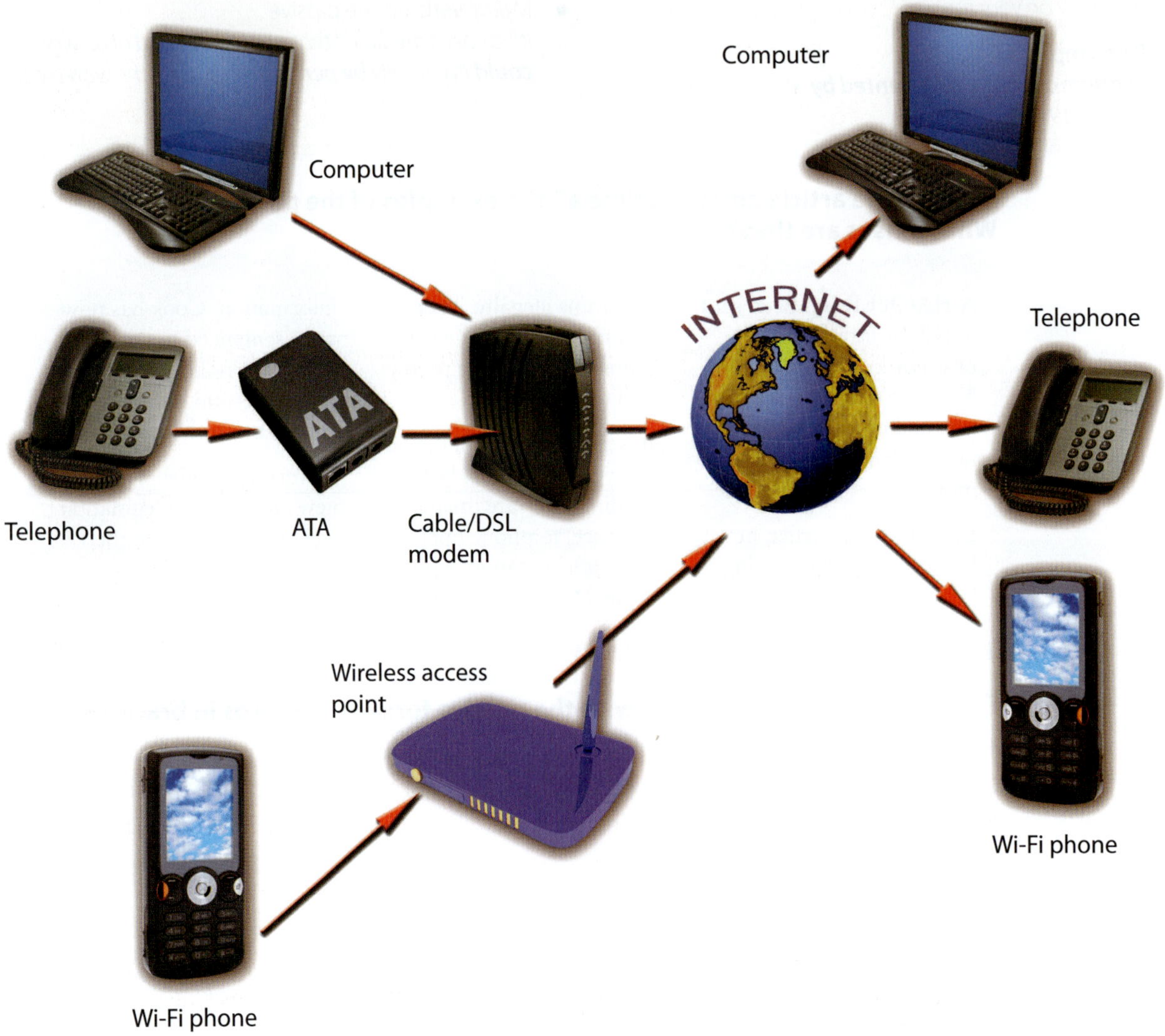

4 Mobile phones

A Label the mobile phone with features from the box.

LCD screen Brand Built-in camera Changeable faceplate SIM card (Subscriber Identity Module) Wireless support Keypad Ringtone

a ....

b ....

c ....

d ....

e ....

f ....

g ....

h ....

B In pairs, describe your mobile phone. Use A and the *Useful language* box to help you.

Useful language

My phone is a … It's got a … With the … , I can …
The best feature is … I never use the … I mostly use it for …

C In pairs, discuss these questions.

1 How much money do you spend on your mobile?
2 Can you send MMS (multimedia messages) from your mobile?
3 Do you access the Internet from your mobile? Which sites do you visit?
4 Can you listen to music and watch TV on your mobile?
5 Do you use your mobile phone for business? Do you think it is secure to carry out financial transactions via mobile phones?
6 Do you ever use your phone while driving?
7 Have you ever had to use your phone in an emergency?
8 Do you think that prolonged use of mobile phones can affect our health (for example cause fatigue and headaches, emit radiation, excite brain cells, etc.)?

An Apple iPhone combines three products – a mobile phone, an iPod, and an internet device with email, web browsing, maps and searching

D Write a summary of the discussion in C as if you were posting it on a blog. Show your summary to other members of your class so that they can add comments.

1 Small networks

A **In pairs, discuss these questions.**

1 What is a computer network?

2 What are the benefits of using networks?

B **Listen to an extract from a lecture on networks and answer these questions.**

1 What does LAN stand for?

2 Where are LANs usually located?

3 What is the difference between a *wired LAN* and a *wireless LAN*.

C **Listen again and label the elements of this LAN.**

(1) A .... wired and wireless LAN

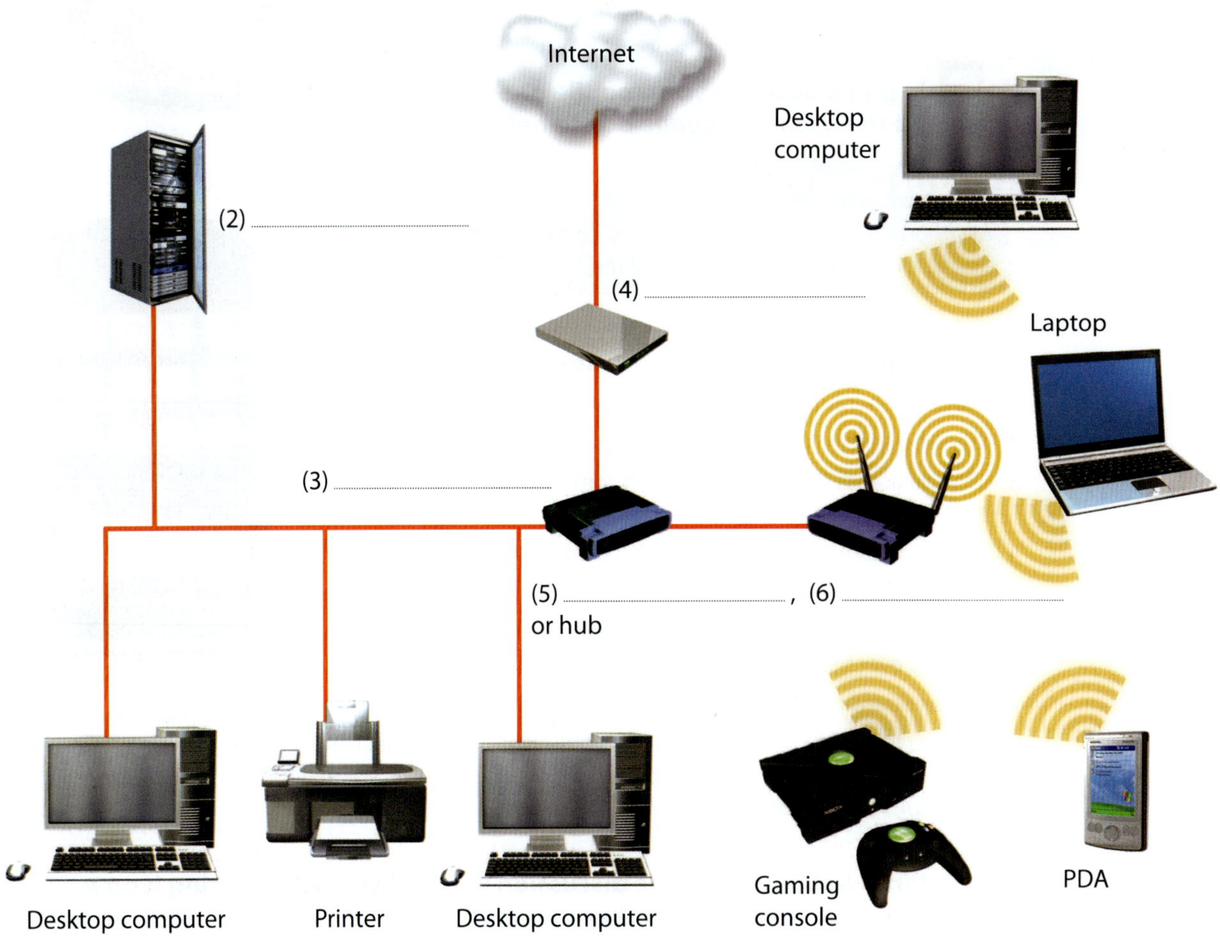

2 Networking FAQs

A Look at the FAQs (i–vi) without reading the whole text. In pairs, try to answer as many of the questions as you can.

B Read the whole text and answer these questions.

1 What does PAN stand for?
2 What is a network protocol?
3 How do you log on to an Internet Service Provider?
4 WiMAX is a type of wireless network. What is it used for?
5 What equipment do you need to set up a wireless LAN?
6 What are the advantages and disadvantages of wireless networks?

Networking FAQs

i How many types of network are there?

Networks are classified according to different criteria:

- **Geographical area**: **PANs** (**P**ersonal **A**rea **N**etworks) typically include a laptop, a mobile phone and a PDA; **LANs** cover a building; **MANs** (**M**etropolitan **A**rea **N**etworks) cover a campus or a city; **WANs** (**W**ide **A**rea **N**etworks) cover a country or a continent.
- **Architecture**: In a **client-server** network, a computer acts as a server and stores and distributes information to the other nodes, or *clients*. In a **peer-to-peer** network, all the computers have the same capabilities – that is, share files and peripherals without requiring a separate server computer.
- **Topology**, or layout: In a **bus** network, all the computers are connected to a main cable, or bus. In a **star** network, all data flows through a central hub, a common connection point for the devices in the network. In a **ring** network, all devices are connected to one another in a continuous loop, or ring.
- **Network protocol**: This is the language, or set of rules, that computers use to communicate with each other. Networks use different protocols. For instance, the Internet uses TCP/IP.

ii How do I install a wired modem router?

A modem **router** is a device that connects your computer or home LAN to the Internet.

- Plug one end of the phone cord directly into a phone jack, and the other end into the ADSL port on the router.
- Plug one end of the Ethernet cable into your computer's network port and the other end into an Ethernet port on the router.
- Turn on your computer. To set up, or configure, the router, you'll need to input some parameters, for example your ISP's name and phone number.

NOTE: A router has various Ethernet ports, so you can connect various PCs to the router via Ethernet cables. If you already have a hub or switch connecting a LAN, you only need one cable to connect the hub to the router.

iii How do I log on to the Internet Service Provider?

You need to type in your username and password. Once you are online, you can get email, look for information on the Web, look up IT words in dictionaries, try out new software, and sign up for RSS feeds, newsletters, etc. It is important that you remember to log off after using the Internet. An open line increases the risk of viruses, and hackers might break into your computer to steal confidential data.

iv What is wireless networking?

Wired networks are linked by Ethernet cables, phone lines and high-speed fibre optic cables. Wireless networks, however, use electromagnetic waves, such as radio waves, to transmit data. These are the main types of wireless networks:

- **Satellites** – for long distances
- **WiMAX** – for connecting Wi-Fi hotspots
- **Wi-Fi** – for medium-range distances
- **Bluetooth** – for short distances
- **GSM** – for mobile phones

v What do I need to set up a home wireless LAN?

You'll need computers equipped with a wireless adapter or wireless card, a wireless access point (a wireless router) and a broadband internet connection.

vi Which is better, a wired or wireless LAN?

Wired LANs are more difficult to install, but they are cheaper, faster and more reliable. Wireless networks let you move, or roam, from one access point to another, but they are less secure and subject to interference.

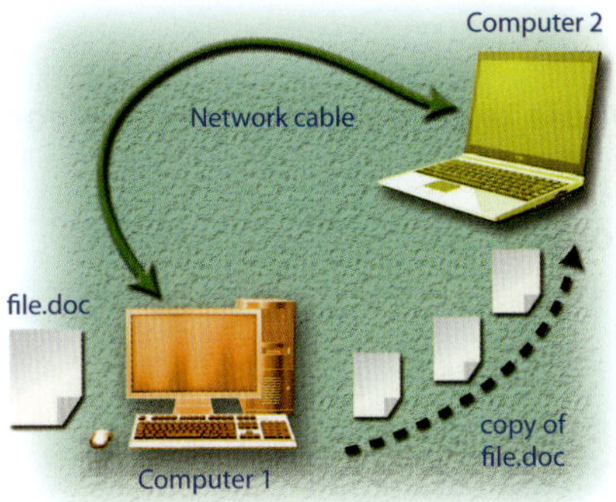

*In a **basic network**, two computers are connected by cable to allow **file sharing**.*

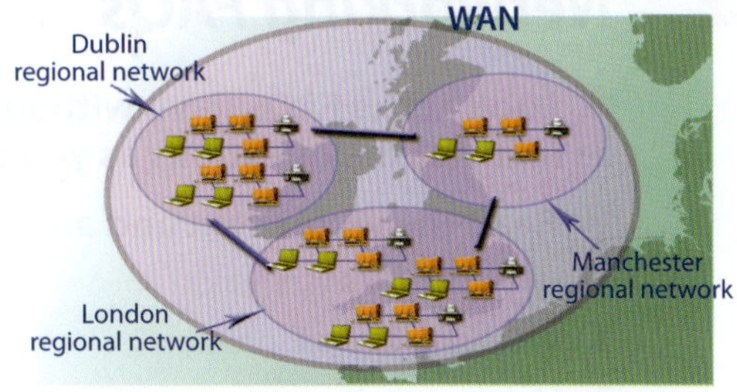

***WANs** cover a large geographic area, like a country or even multiple countries. They are built by large telecommunication companies. The largest WAN in existence is the Internet.*

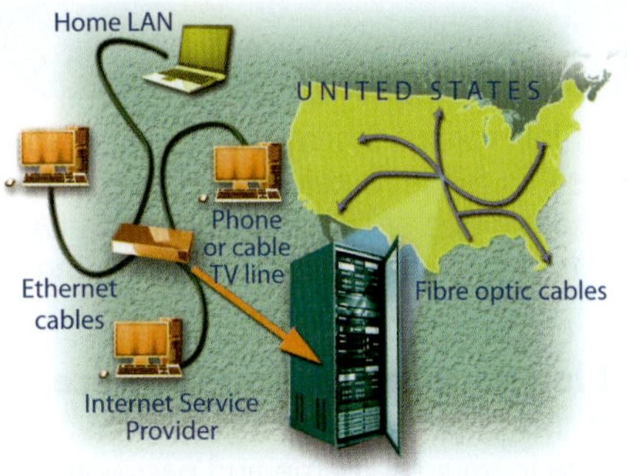

*In many homes, **Ethernet cables** are used to connect computers. **Phone** or **cable TV lines** then connect the home LAN to the ISP. Much of the Internet uses high-speed **fibre optic cable** to send data over long distances.*

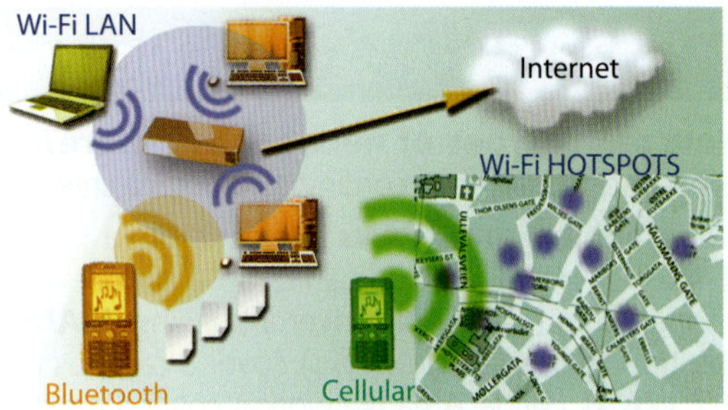

***Wi-Fi** is the standard technology for building wireless LANS and public **hotspots**. **Bluetooth networks** allow handhelds, mobile phones and other devices to communicate over short distances. **Cellular networks** are used in mobile phone communications.*

C In pairs, do this network quiz. See which pair can finish first.

1 This network typically consists of two or more local area networks, covering a large geographical area.

a LAN **b** WAN **c** Intranet

2 This type of network does not have a dedicated server; all the computers are independent.

a peer-to-peer **b** client-server **c** Metropolitan Area Network

3 On this topology, all devices are connected to the same circuit, forming a continuous loop.

a star **b** ring **c** bus

4 The language used by computers to communicate with each other on the Internet is called

a Ethernet. **b** ADSL. **c** TCP/IP.

5 Which cables are used to transfer information for the Internet over long distances at high speeds?

a telephone lines **b** Ethernet cables **c** fibre optic cables

6 Which device allows several computers on a local network to share an internet connection?

a an ADSL port **b** a router **c** an Ethernet port

7 Which device serves as a common connection point for devices in a wireless network?

a wireless access point **b** wired router **c** wireless adapter

8 Bluetooth is a wireless technology that uses radio waves to transmit data over

a long distances. **b** medium-range distances. **c** short distances (ten metres or less).

3 Language work: phrasal verbs

A Look at the HELP box. Do you have the equivalent of phrasal verbs in your language? How do you say the phrasal verbs in the HELP box?

HELP box

Phrasal verbs

- The meaning of some verbs with particle (often called phrasal verbs) can be easily understood from its two parts.

  Look at *the photos.*

  A network ***consists of*** *two or more …*

  Separate networks are ***linked over*** *a public network, the Internet.*

- However, many phrasal verbs have an idiomatic meaning, not predictable from the meaning of its parts.

  carry (= transport); **carry out** (= execute)

  Computers ***carry out*** *the programs …*

- Certain particles have similar meanings, regardless of the verb (**on**/**off**, **in**/**out**, etc.).

  turn on / **switch on**
  (= start the operation of something)
  turn off / **switch off**
  (= stop the operation of something)

- Other common phrasal verbs in computing include:

  plug into (= connect)
  Plug *one end of the phone cord* ***into*** *the phone jack.*
  set up (= establish)
  What do I need to ***set up*** *a wireless LAN?*
  sign up (= register, enrol in a service)
  Once connected, you can ***sign up*** *for RSS feeds, newsletters, etc.*
  try out (= test or use experimentally)
  You can ***try out*** *new software on their site.*
  find out (= learn, discover)
  Search the Web to ***find out*** *more information about WiMAX.*
  take up (= occupy)
  Fibre optic cables ***take up*** *less space than copper cables.*
  make up (= constitute, form)
  Several LANs connected together ***make up*** *a WAN.*
  fill in (= write the necessary information)
  You need to ***fill in*** *this online form.*

- When the verb has a preposition associated with it, the preposition must precede the object:

  You can ***look for*** *information on the Web.*
  (**not**: ~~***look*** *information* ***for***~~)

  Hackers might ***break into*** *your PC.*
  (**not**: ~~***break*** *your PC* ***into***~~)

  When the particle is an adverb, it can precede or follow the direct object:

  You need to ***type in*** *your username /*
  … ***type*** *your username* ***in****.*
  You can ***look up*** *words in a dictionary /*
  … ***look*** *words* ***up*** *in a dictionary.*
  Turn on *the computer. /*
  Turn *the computer* ***on****.*

  If the direct object is a pronoun, the: particle must follow it

  You need to ***type it in****.*
  (**not**: ~~***type in*** *it*~~)

B Complete these sentences with the correct form of a phrasal verb from the HELP box.

1 To join the club, ... this form and send it to our office.
2 The CPU .... all the basic operations on the data.
3 Digital music .... a lot of space – about 10 MB for every minute of stereo sound.
4 Thousands of networks .... the Internet.
5 You can use newsgroups to .... about the latest trends, customer needs, etc.

C Match the questions (1–6) with the answers (a–f).

1 Why was the hacker arrested?
2 Is it OK to **log on** to my bank account using public computers in a cybercafé?
3 How do I **set up** an internet connection at home?
4 Can I download software from your site?
5 How can I add video to instant messaging?
6 What do I need to do to **sign up** for a *Yahoo!* email account?

a Yes, but always remember to **log off** after you've ended your session.
b Yes, you can even **try** the programs **out** for a period before you buy them!
c Because he **broke into** a computer system and stole confidential data.
d Simply install this program and **plug** the webcam **into** your computer.
e You need to install the software for your router. Follow the instructions provided by your ISP, probably in the form of a .pdf file on a CD.
f You have to create a username and password and then give some personal details.

4 *WANs and satellites*

A Prepare a description of the network below to present to the rest of the class. Use PowerPoint if possible. Use the *Useful language* box, the HELP box on page 143 and the text on pages 141–142 to help you.

Useful language

The diagram represents/shows …
This network is made up of / consists of …
Two networks are connected via …
The computers are linked up to …
The satellite receives signals from …
The signals are sent on to …
The purpose of … is to …

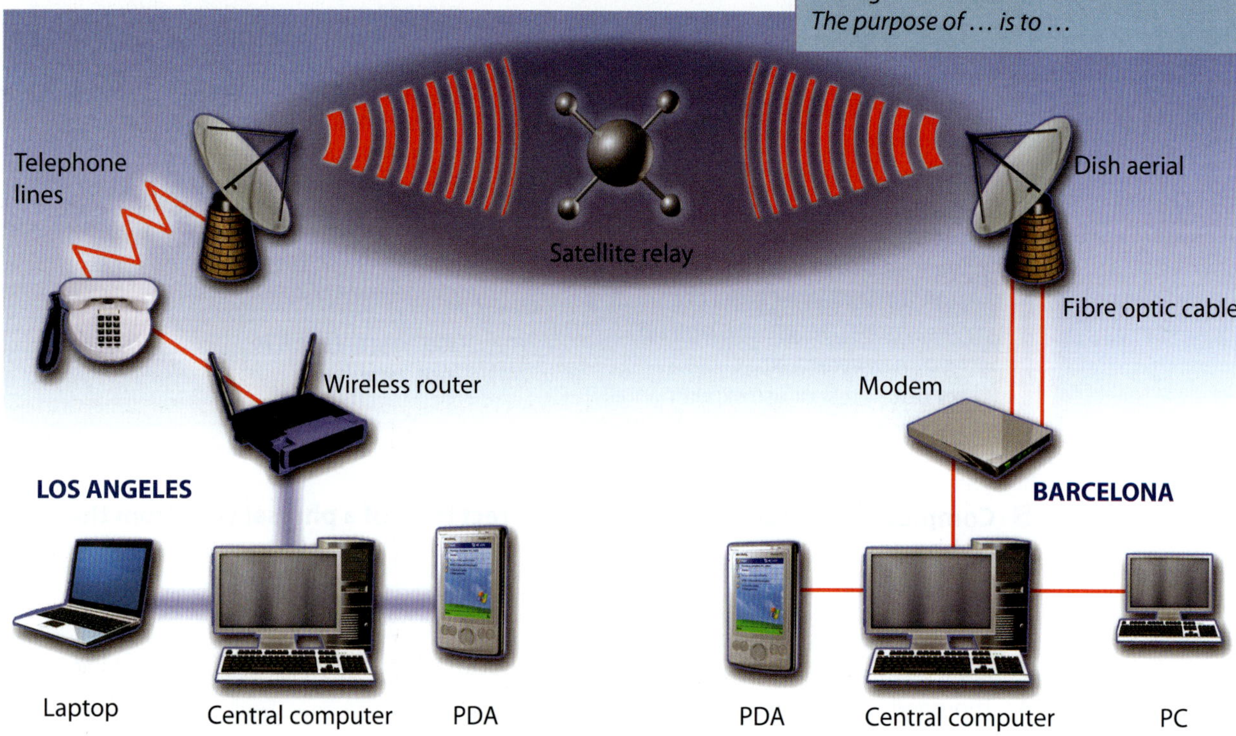

B Present your description to the rest of the class.

Unit 29 Video games

1 Game platforms

A In pairs, discuss these questions.

1 Do you play video games?

2 What are your favourite games? Make a list.

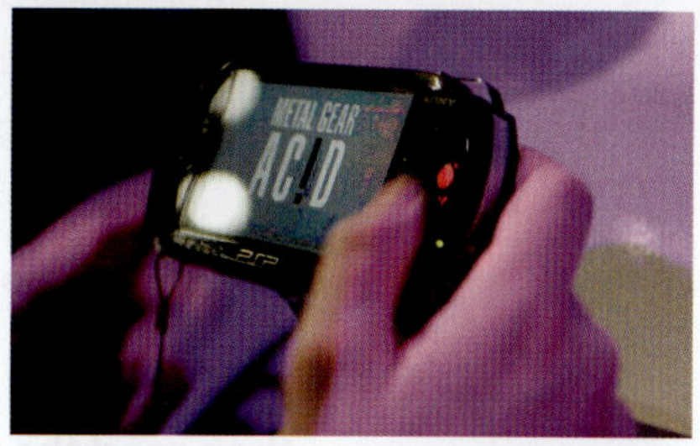

a ....

B Label the pictures (a–f) with the types of game in the box.

PC games   Console games   Arcade games
Handheld games   Mobile phone games
Massively multiplayer online games

b ....

c ....

C Video games are played on a variety of electronic devices, or platforms. Complete these sentences with game platforms from the box and types of game from B.

Personal computer   Video game consoles
Portable gaming devices   3G mobile phones

1 .... are played on ...., such as the Sony PS3 or Microsoft Xbox 360. In the past, these electronic devices were just connected to a standard TV or video monitor; now they can also be connected to the Net, via cables or wirelessly.

2 .... are played on ...., such as the Sony PSP and the Nintendo DS. You can also play games on some graphing calculators and watches.

3 Don't worry if you don't have a game console. You can still play .... on a .... The graphics are even more impressive if you have a high-resolution monitor. You can buy games on CDs and DVDs, or download them from the Internet.

4 .... allow you to play against other users in other parts of the world using the Internet – something unique to electronic gaming. Players connect to a game server hosted by an ISP, a game company, or an individual enthusiast.

5 Some .... are programmed to run natively on the chip of .... For instance, *Snake* is installed on many Nokia phones. Many Java-based games are also available via download.

6 .... are played on coin-operated machines, typically installed in restaurants, bars and amusement arcades. For example, you can fly an aircraft or a spaceship using a joystick.

d ....

e ....

f ....

D **In pairs, discuss these questions. Give reasons for your answers.**

1 Which is your favourite game platform? What advantages and disadvantages does it have over other game platforms?

2 Which game platform would you most like to own?

3 Do you play games on your mobile phone? What is the experience like?

2 *Game genres*

A How many different game genres can you think of? In pairs, make a list and then read the text to see how many genres from your list are mentioned.

Game genres

There are so many different genres and mixes of genres that it's difficult to put each game into a specific category. In the following article we'll cover the basic genres that differentiate between games.

The **First-person shooter** (**FPS**) and **Action** genres are currently the most popular. Games like *Half-Life*, *Halo* and *Call of Duty* are the most popular games in the FPS category. For Action, innovative titles like the *Grand Theft Auto* series, *Gears of War* and *Splinter Cell* are huge successes.

The **Role-playing game** (**RPG**) genre has remained strong throughout the entire history of console and PC gaming. Current hits like *Final Fantasy XII*, *Oblivion* and the *Knights of the Old Republic* series are all based on RPG roots. The recent development of *massively multiplayer online RPGs* has been made possible by widespread broadband access, allowing gamers to play internationally with thousands of people across the globe in a constant virtual world.

Adventure games and **Puzzle games** remain strong despite being limited in scope and technology. The new concept of *party* games – where people play together in multiplayer mode – has recently injected new life into this genre. Titles like *Zelda* and *Wario-Ware* are familiar names.

Sports games are an increasingly popular portion of the gaming industry. Electronic Arts (EA) have been making games licensed from the NBA, NFL and MLB for over a decade. Another sector of the Sports industry is the entire racing sub-genre. Massive hits like the *Burnout* and *Need for Speed* series are hugely exciting, and the crashes can be realistic and terrifying.

Halo 3 is very popular on the Xbox console; millions of people also play the game online

The **Simulation** genre has enjoyed wild success, including the best-selling PC games of all time: *The Sims* & *The Sims 2*. The entire *Sims* series, designed by Maxis, is dominant in this genre. Jet fighter and flying sims are also important types of simulation game.

Strategy is a genre mainly restricted to PC, largely because the mouse and keyboard are central to gameplay. There are a few good Strategy games for console, however. Big names in Strategy include *Warcraft III*, *Starcraft*, *Command and Conquer* and *Warhammer 40,000*.

Finally, we have the **Fighting** genre. Developed from early hit games like *Street Fighter II*, Fighting games have enjoyed a renaissance as they've been updated fully to include 3-D characters and arenas. Titles like *Dead or Alive*, *Tekken* and *Soul Calibur* are big favourites.

So what kind of game player are you? Chances are that if you're a PC gamer, you prefer FPS, RPG, Simulation, and Strategy games. The console gamer typically enjoys Sports, Racing, Fighting, RPGs, and a few FPS titles. Of course, many people own both a console and a PC, therefore combining the best of both worlds.

B These statements about gaming are all false. Read the text again and correct them.

1 Role-playing games are currently the most popular.
2 Massively multiplayer online RPGs have been made possible by widespread internet access.
3 *Oblivion* is an Action game.
4 *The Sims* series is the least popular in the Simulation category.
5 Strategy games are mainly restricted to game consoles.
6 *Warcraft* belongs to the Fighting genre.
7 Console gamers typically prefer Simulation and Strategy games.

C Find words or phrases in the text with the following meanings.

1 now; at this time or period (lines 5–10) ....
2 existing or happening in many places and/or among many people (lines 15–20) ....
3 in spite of; notwithstanding (lines 20–25) ....
4 more and more (lines 25–30) ....
5 a smaller category within a particular genre (lines 30–35) ....
6 big successes (lines 30–35) ....
7 sold in very large numbers (lines 35–40) ....
8 modernized (lines 45–50) ....

D In pairs, discuss these questions. Give reasons for your answers.

1 What is your favourite and least favourite genre of game?
2 What are your favourite games? Describe them to your partner.

3 *Language work: adverbs*

A Look at the HELP box on page 148 and then complete these sentences with the adverbial form of the words in brackets.

1 Simulation games are (wide) .... used in both universities and businesses.
2 Massively multiplayer online RPGs have (recent) .... become more popular, mainly due to faster internet connections.
3 Strategy is a genre (main) .... restricted to PC.
4 Video games often come with a clear set of motivation tools, such as scores and moving to higher levels when a player performs (good) ....
5 Cheap PCs don't process data (fast) .... enough to support high-end games.

B Are the words in bold adjectives or adverbs? Write *adj* or *adv*.

1 Atari's platform was the most popular **early** video game console, and many developers emulated Atari games to attract customers. ....
2 The chess game ended **early**, at the 24th move. ....
3 On the TPS Report gaming blog, you will find reviews, a forum and a **monthly** podcast. ....
4 The podcast is broadcast **monthly**. ....
5 You have to work **hard** to succeed in the gaming industry. ....
6 Some experts say that **hard** work makes people happy. ....

HELP box

Adverbs

- We use adverbs to give information about an action. Adverbs of manner, time and place describe how, when or where something happens.

  They've been updated ***fully*** *to include 3-D characters.* (= manner, i.e. *how*)

  The Action genre of games is ***currently*** *the most popular.* (= time, i.e. *when*)

  ... allowing gamers to play ***internationally*** *...* (= place, i.e. *where*)

  We also use adverbs to modify adjectives.

  Sports games are an ***increasingly popular*** *portion of the gaming industry.*

- We usually form an adverb by adding **-ly** to an adjective.

  typical → *typical**ly***
  The console gamer ***typically*** *enjoys Sports, ...*

- With adjectives ending in **-y**, we change the **y** to **i** before adding the ending **-ly**.

  easy → *eas**ily***
  The Nintendo Wii connects ***easily*** *to the Internet.*

- Note that not all words that end in **-ly** are adverbs. These words are adjectives: **friendly**, **deadly**, **lovely**, **lonely**.

- The adverb from **good** is **well**.

  His French is very ***good****. He speaks French* ***well****.*

- Some words have the same form as an adjective and an adverb (e.g. **fast**, **hard**, **early**, **late**, **daily**, **monthly**).

  New games require a ***fast*** *processor.* (= adjective)
  The processor speed tells you how ***fast*** *your PC executes instructions.* (= adverb)

4 Present and future trends in gaming

A Listen to an interview with Matt Robinson, the administrator of the TPS Report gaming blog. How many game platforms does he mention?

B These statements about video games are all false. Listen to the interview again and correct them.

1 Video games are popular because they are fun and addictive.
2 Well-known Hollywood actors appear in video games.
3 The Nintendo Wii is aimed at hardcore gamers.
4 It's free to play *World of Warcraft*.
5 Holography is an advanced form of photography that uses lasers to produce two-dimensional images.
6 In the future, gesture recognition systems will produce photo-realistic images.

C **Listen again and complete these extracts from the interview with adverbs.**

1 With a game you are .... in control of the action.
2 Games are now even more life-like and .... attractive.
3 A lot of modern games .... draw inspiration from films and even TV.
4 Their .... released Wii console has an inexpensive, simplistic, *pick up 'n' play* feel to it.
5 Wii is .... the most popular of the three machines.
6 Logging onto an .... separate universe to meet and play alongside your friends has enormous attraction.
7 Mobile gaming has .... been about easy, simplistic 2-D games.

D **In pairs, discuss if you agree with everything that Matt says in his interview about the future of gaming. What are your own predictions?**

5 *The pros and cons of gaming*

A **In pairs, look at the statements about gaming (1–4) and say if you agree or disagree with them. Give reasons for your answers.**

1 TV and video games are amusing and can be educational. But too much of this kind of entertainment can be addictive and make children become accustomed to violence.
2 Massively multiplayer online games are interactive and fun.
3 Video games have negative effects on children and distract them from school and homework.
4 Modern games and simulations offer a great deal of adventure and challenge. In addition, they can teach skills such as strategic thinking, interpretative analysis and problem solving.

B **Write an essay called *The pros and cons of gaming* (80–120 words). Use these steps and the *Useful language* box to help you.**

- The **opening** (paragraph one):
  Present the topic in one or two sentences.
- The **body** (paragraphs two and three):
  Give pros (arguments in favour) with facts and examples. Give cons (arguments against) with facts and examples.
- The **closing** (paragraph four):
  Summarize your main ideas and give your opinion.

Useful language

To add arguments:
In addition, … *Furthermore, …*

To introduce opposing ideas:
On the one hand, … *On the other hand, …*
Some people say … *Others say …* *However, …*

To express opinions:
In my opinion, … *I believe that …*
It seems to me that … *It's clear that …*

To conclude:
In conclusion, … *To sum up, …* *In short, …*

1 Future trends

A **In pairs, discuss these questions.**

1 What do you think a *trend* is?

2 What trends in ICT do you think will affect our lives in the future? Make a list.

B **Match the texts (1–5) with the pictures (a–e). Which trends from your list in A are mentioned?**

a ....

b ....

c ....

d ....

e ....

1

By all accounts, **nanotechnology** – the science of making devices from single atoms and molecules – is going to have a huge impact on both business and our daily lives. Nano devices are measured in **nanometres** (one billionth of a metre) and are expected to be used in the following areas.

- **Nanocomputers**: Chip makers will make tiny microprocessors with **nanotransistors**, ranging from 60 to 5 nanometres in size.
- **Nanomedicine**: By 2020, scientists believe that nano-sized robots, or **nanobots**, will be injected into the body's bloodstream to treat diseases at the cellular level.
- **Nanomaterials**: New materials will be made from carbon atoms in the form of **nanotubes**, which are more flexible, resistant and durable than steel or aluminium. They will be incorporated into all kinds of products, for example stain-resistant coatings for clothes and scratch-resistant paints for cars.

2

Artificial Intelligence (**AI**) is the science of making intelligent machines and programs. The term originated in the 1940s, when Alan Turing said: 'A machine has artificial intelligence when there is no discernible difference between the conversation generated by the machine and that of an intelligent person.' A typical AI application is **robotics**. One example is ASIMO, Honda's intelligent humanoid robot. Soon, engineers will have built different types of **android**, with the form and capabilities of humans. Another AI application is **expert systems** – programs containing everything that an 'expert' knows about a subject. In a few years, doctors will be using expert systems to diagnose illnesses.

3

Imagine you are about to take a holiday in Europe. You walk out to the garage and talk to your car. Recognizing your voice, the car's doors unlock. On the way to the airport, you stop at an ATM. A camera mounted on the bank machine looks you in the eye, recognizes the pattern of your iris and allows you to withdraw cash from your account.

When you enter the airport, a hidden camera compares the digitized image of your face to that of suspected criminals. At the immigration checkpoint, you swipe a card and place your hand on a small metal surface. The geometry of your hand matches the code on the card, and the gate opens. You're on your way.

Does it sound futuristic? Well, the future is here. **Biometrics** uses computer technology to identify people based on physical characteristics such as fingerprints, facial features, voice, iris and retina patterns.

Adapted from the *Richmond Times-Dispatch*

4

Ubiquitous computing, also known as **pervasive computing**, is a new approach in which computer functions are integrated into everyday life, often in an invisible way. **Ubiquitous devices** can be anything from smartphones to tiny sensors in homes, offices and cars, connected to networks, which allow information to be accessed anytime and anywhere – in other words, ubiquitously. In the future people will interact naturally with hundreds of these **smart devices** (objects containing a microchip and memory) every day, each invisibly **embedded** in our environment and communicating with each other without cables.

5

In the ideal **smart home**, **appliances** and electronic devices work in sync to keep the house secure. For example, when a regular alarm system senses that someone is breaking into the house, it usually alerts the alarm company and then the police. A smart home system would go further, turning on the lights in the home and then sending a text message to the owner's phone. Motorola *Homesight* even sends images captured by wireless cameras to phones and PCs.

Smart homes can remember your living patterns, so if you like to listen to some classical music when you come home from work, your house can do that for you automatically. They will also know when the house is empty and make sure all appliances are turned off. All home devices will be interconnected over a home area network where phones, cable services, home cinemas, touch screens, smart mirrors and even the refrigerator will cooperate to make our lives more comfortable.

Adapted from ***www.businessweek.com***

C Read the texts again and answer these questions.

1 Which unit of measurement is used in nanotechnology?
2 What are the advantages of nanotubes over regular materials?
3 What will doctors use *expert systems* for?
4 What features are analysed by biometrics?
5 Which trend refers to computers embedded in everyday devices, communicating with each other over wireless networks?
6 What will the alarm system do if someone breaks into a smart home?
7 How will devices be interconnected inside the smart home?

D Find words in the texts with the following meanings.

1 a microscopic robot, built with nanotechnology (text 1) ....
2 a robot that resembles a human (text 2) ....
3 biological identification of a person (text 3) ....
4 integrated; inserted into (text 4) ....
5 electrical devices, or machines, used in the home (text 5) ....

E Write a suitable caption for each picture on page 150.

2 *RFID tags*

A **Listen to Sarah Wood, an ICT teacher, giving a class about RFID tags. Which definition (a–c) best describes RFID?**

a a smart technology worn on the user's body so that they can email and access the Web

b a technology that uses radio waves and chip-equipped tags to automatically identify people or things

c a technology that uses microchips and bar codes to track people or things at a distance

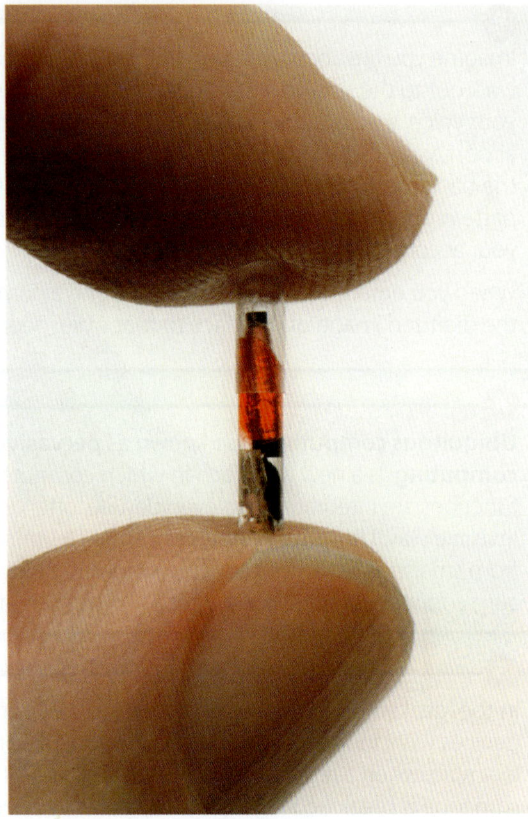

An RFID micro chip

B **Listen again and decide which answers (a or b) are correct.**

1 RFID stands for
 - **a** **R**adio **F**requency **Id**entification.
 - **b** **R**adio **F**requency **I**dentification **D**ownload.

2 Radio tags
 - **a** can only be attached to or embedded into products.
 - **b** can be attached to or embedded into products, animals and humans.

3 Active RFID tags
 - **a** have a communication range of several hundred metres.
 - **b** have a communication range of five metres.

4 RFID chips
 - **a** will help us track ordinary objects like car keys or books.
 - **b** won't be able to locate objects when they are lost or stolen.

5 Radio tags may be implanted under the skin
 - **a** to confirm a patient's identity and cure illnesses.
 - **b** to give doctors instant access to a patient's medical history.

6 According to consumer organizations, RFID tags
 - **a** could be used to track consumers or to steal a person's identity.
 - **b** are secure and private; there is no need for concern.

C **In pairs, discuss how secure you think RFID is. Do you agree with the consumer organizations or the manufacturers? Give reasons for your answers.**

3 Language work: future forms

A Look at the HELP box and then choose the correct words in brackets to complete these sentences.

1 In the future, I hope we ('ll have / 're going to have) robots in the home to help us with the housework.

2 Hey, Nick, be careful, you ('re going to spill / 'll spill) that coffee on the computer!

3 It's John's birthday next week. We ('ll give / 're going to give) him a mobile phone.

4 – My laptop has crashed!
– Don't worry. I ('ll lend / 'm going to lend) you mine.

5 The Internet (will probably change / is probably going to change) the publishing industry in the way that TV changed the movie industry.

6 Futurists predict that smart technology (will be / is going to be) incorporated into fabrics, so you'll be able to email from your coat!

HELP box

Future forms

We use the future simple (**will/won't** + verb) in the following ways:

- To make predictions when you don't have present evidence that something will happen

  *Nanobots **will be injected** into the body's bloodstream to treat diseases.*

- To talk about hopes and promises, especially with the words **expect**, **think**, **hope** and **probably**

  *They hope that people **will interact** naturally with hundreds of smart devices at a time.*

- To describe an instant decision, often when we make an offer

  *Sure, **I'll help** you with your homework.*

- To talk about facts that will inevitably happen

  ***She'll be** 21 in May.*

We use **be going to** + verb in the following ways:

- To describe future intentions

  *She's **going to write** a book about ubiquitous computing.*

- To make predictions when you have present evidence that something is going to happen

  *By all accounts, nanotechnology **is going to have** a huge impact on business and our daily lives.*

We use the future continuous (**will be** + **-ing** form of the verb) to talk about actions in progress at a specific time in the future.

*In a few years, doctors **will be using** expert systems to diagnose illnesses.*

We use the future perfect (**will have** + past participle) to talk about actions finished at a specific time in the future.

*Soon, engineers **will have built** different types of android.*

B Complete these sentences with the correct future form of the verb in brackets. Use the future continuous or future perfect.

1 Thanks to ICT, by the year 2030 we (find) .... cures for the major diseases of our time.

2 In twenty years' time, some people (live) .... in space, perhaps inside a computerized colony.

3 By this time next week, I (work) .... for IBM.

4 By this time next month, I (buy) .... that BlackBerry that I've been wanting to buy for months.

5 Scientists predict that in twenty years' time nearly everyone (live) .... in smart houses.

C **In pairs, discuss these predictions. Do you agree or disagree? Give reasons for your answers. Look at the *Useful language* box to help you.**

1 Some day, we'll be talking to computers naturally, like friends.
2 Microchips implanted in our arms will serve as ID cards and contain our medical records.
3 Robots will learn to build themselves, without human help.
4 Smart homes will be voice-activated.
5 Computers will be ubiquitous and almost invisible, embedded into our homes and integrated into our lives.

Useful language

I think that ...
What do you think about number ... ?
I'm not sure that ... will ...
I completely agree/disagree with ...

4 *Making predictions*

A **Write your own predictions about these topics.**

- **Work/Jobs**
  Example: *By the year 2030, human labour in industry will have been replaced by robots.*
  Your prediction: ....
- **Money**
  Example: *Cash will be replaced by electronic money.*
  Your prediction: ....
- **Education**
  Example: *By the end of this century, every student in every school will have a PC.*
  Your prediction: ....
- **The Internet**
  Example: *People in every country will have high-speed access to the Internet within five years.*
  Your prediction: ....

B **In pairs, compare your predictions. Find out more about your partner's predictions.**

Now visit www.cambridge.org/elt/ict for an online task.

Appendix: a model CV

Curriculum vitae

Personal information

Name: María Quintana

Address: Avda Séneca, 5, Madrid 28040

Telephone: 00 34 91 5435201

Email: mquintana0782@telefonica.net

Date of birth: 28/07/82

Education and Training

2006	Online diploma in web-based technology for business, www.elearnbusiness.com
2005	Course in web design at the Cybernetics College, London: HTML, Java and Macromedia Dreamweaver
2004	Course in computer hardware and networking at the Cybernetics College, London
1999–2004	Degree in Computer Science and Engineering, University of Madrid

Work experience

January 2006 – present	Part-time Webmaster at www.keo.es; responsible for updating the site and using Adobe Flash to create animations
May 2005 – December 2006	IT consultant at Media Market, specializing in e-commerce and IT strategies

IT skills

Knowledge of multiple computer platforms (Windows, Mac and Linux); strong database skills (including the popular open source MySQL database); complete understanding of graphics formats and Cascading Style Sheets

Personal skills

Social and organizational skills

Good communication skills

Languages

Spanish mother tongue; English (Cambridge CAE); Arabic (fluent)

Hobbies and Interests

Web surfing, listening to music and travelling

References

Miguel Santana, Manager, keo.es

Sam Jakes, Lecturer, Cybernetics College

Glossary

A

@ /ət/ The 'at' sign that separates the recipient's name from the domain name in an email address.

Acrobat Reader /ˌækrəbæt 'riːdə/ *n* Adobe's free software for displaying and printing PDF files.

ADSL /eɪdiːes'el/ *n* Asymmetric Digital Subscriber Line. A broadband communication technology designed for use on telephone lines; it allows a single phone connection to be used for both internet service and voice calls at the same time.

ADSL modem /ˌeɪdiːes el 'məʊdem/ *n* A device used to connect one or more computers to an ADSL phone line.

adware /'ædweə/ *n* Software devised to display advertisements; some may include spyware.

alphanumeric keys /ˌælfənjuːmerɪk 'kiːz/ *n* Keys that represent letters and numbers, arranged as on a typewriter.

Amazon.com /'æməzən ˌdɒt ˌkɒm/ *n* A popular online shopping site.

android /'ændrɔɪd/ *n* A robot that resembles a human.

animations /ænɪ'meɪʃənz/ *n* Images made up of a series of independent pictures put together in sequence to look like moving pictures.

applet /'æplət/ *n* A small Java application, usually designed to run automatically within a web page.

application software /æplɪ'keɪʃən ˌsɒfweə/ *n* Programs that let you do specific tasks, such as word processing, database management or financial planning.

arcade game /ɑː'keɪd ˌgeɪm/ *n* A game played in a coin-operated machine, typically installed in pubs, restaurants and amusement arcades.

arithmetic logic unit (ALU) /əˌrɪθmətɪk 'lɒdʒɪk ˌjuːnɪt/ n A component of the CPU which performs the actual arithmetic and logical operations asked for by a program.

ARPANet /'ɑːpənet/ *n* Advanced Research Projects Agency Network. Developed in the early 70s by the US Department of Defence. This was the precursor to the Internet.

Artificial Intelligence /ɑːtɪˌfɪʃəl ɪn'telɪdʒəns/ *n* The study of methods by which a computer can simulate aspects of human intelligence.

ASCII code /'æskiː ˌkəʊd/ *n* A standard system for the binary representation of characters. ASCII, which stands for American Standard Code for Information Interchange, permits computers from different manufacturers to exchange data.

aspect ratio /'æspekt ˌreɪʃiəʊ/ *n* The width of the screen divided by its height, e.g. 4:3 (standard PC monitor or TV set) and 16:9 (high-definition TV).

assembler /ə'semblə/ *n* A special program that converts a program written in a low-level language into machine code.

assembly language /ə'sembli ˌlæŋgwɪdʒ/ *n* A low-level language that uses abbreviations, such as ADD, SUB and MPY, to represent instructions.

ATA adaptor /'eɪtiːeɪ əˌdæptə/ *n* An analogue telephone adaptor, which converts the analogue signals of your traditional phone into digital signals.

Athlon /'æθlɒn/ *n* A processor manufactured by AMD.

attachment /ə'tætʃmənt/ *n* A file that has been included as part of an email message.

attributes /'ætrɪbjuːts/ *n* Characteristics that affect the visual representation of lines and polygons, e.g. line styles, rectangle colour, etc.

authentication /ɔːˌθentɪ'keɪʃən/ *n* Verifying the identity of a user logging onto a network; ways of authentication include passwords, digital certificates and biometrics.

avatar /'ævətɑː/ *n* An object which represents a participant in a 3-D chat room.

B

back up /ˌbæk 'ʌp/ *v* To copy files from one disk to another.

backbone /'bækbəʊn/*n* High-speed lines or connections that form the major access pathways within the Internet.

backup /'bækʌp/ *n* A copy of data or software, usually kept in case the original disk is damaged.

bandwidth /'bændwɪtθ/ *n* The quantity of data that can be transmitted through a network, measured in bits per second (bps).

bar code reader /'bɑːkəʊd ˌriːdə/ *n* A specialized scanner used to read price labels in shops.

BASIC /'beɪsɪk/ *n* A high-level programming language developed in the 1960s, widely used in programming because it was interactive and easy to use. Short for *Beginner's All-purpose Symbolic Instruction Code.*

binary code /'baɪnəri ˌkəʊd/ *n* A code made of just two numbers (0 and 1).

binary digit /'baɪnəri ˌdɪdʒɪt/ *n* The smallest unit of information in the binary system, 0 or 1. Also called a *bit*.

binary system /'baɪnəri ˌsɪstəm/ n A notation system in which numbers are represented by the two digits: 0 and 1.Thus the binary number 10 represents 2 in the decimal system, while 100 represents 4.

biometrics /baɪəʊ'metrɪks/ *n* The science that uses computer technology to identify people based on physical features, such as fingerprints or voice scans. From the ancient Greek: bios = life, metron = measure.

bit /bɪt/ n See **binary digit**.

bit-mapped graphics /ˌbɪtmæpt 'græfɪks/ *n* See **raster graphics**.

BitTorrent /ˌbɪt 'tɒrent/ *n* A protocol for transferring music, films, games and podcasts. It breaks files into chunks and distributes them among a large number of users; when you download a torrent, you are also uploading it to another user.

BlackBerry /'blækbəri/ *n* A wireless handheld device, developed by Research in Motion, which provides email, phone, text messaging, web browsing, an organizer, as well as instant messaging and corporate data access.

blind carbon copy (Bcc) /ˌblaɪnd ˌkɑːbən 'kɒpi/ *n* Addresses in the Bcc: line of an email program will receive a copy of the message, but the identity of the recipients will be kept secret.

blog /blɒg/ **1** n A user-generated website where people express their opinions. The entries are displayed in a reverse chronological order. The term comes from *web log*, coined by Jorn Barge in 1997 to refer to an online diary. **2** *v* To write entries in a blog.

blog administrator /ˌblɒg əd'mɪnɪstreɪtə/ *n* Someone who edits and deletes posts or comments made by contributors to a blog.

blogger /'blɒgə/ *n* A person who writes on a blog.

blogging /'blɒgɪŋ/ *n* Writing web logs. See **blog**.

blogosphere /'blɒgəʊsfɪə/ *n* The collective term including all blogs as a community.

Bluetooth /'bluːtuːθ/ *n* A wireless technology that allows handhelds, mobile phones and other peripheral devices to communicate over short distances.

Blu-ray disc /'bluːreɪ ˌdɪsk/ *n* A new optical disc, created by Sony, which can record and play back high-definition TV and computer data. Unlike current DVDs, which use a red laser to read and write data, Blu-ray uses a blue-violet laser, hence its name.

bookmark /'bʊkmɑːk/ *n* A saved link that takes users directly to a web address. Bookmarks are also called *favourites*.

Braille /breɪl/ *n* A system of writing devised by Louis Braille for blind people, in which combinations of raised dots representing letters and numbers can be identified by touch.

Braille embosser /'breɪl ɪmˌbɒsə/ *n* An impact printer that produces tactile Braille symbols.

brightness /'braɪtnəs/ *n* The amount of light produced by an LCD monitor, measured in cd/m2.

broadband /'brɔːdbænd/ *n* High-speed transmission, usually referring to internet access via cable and ADSL; about 400 times faster than dial-up access.

browser /'braʊzə/ n A program designed to fetch and display web pages on the Internet.

buddy list /'bʌdi ˌlɪst/ *n* A list of people that you may want to communicate with via instant messaging.

bug /bʌg/ *n* An error in a computer program.

built-in /'bɪltɪn/ *adj* Integrated; constructed as part of a larger unit.

bulletin board system (BBS) /ˌbʊlətɪn ˌbɔːd 'sɪstəm/ *n* A system that enables its users, usually members of a particular interest group, to share information and programs.

burn /bɜːn/ *v* To write data to a CD or DVD.

bus /bʌs/ *n* An electrical channel, or highway, which carries signals between units inside the computer.

bus topology /ˌbʌs tɒ'pɒlədʒi/ *n* One of the three principal topologies for a LAN, in which all computer devices are connected to a main cable, or bus.

bus width /'bʌs ˌwɪtθ/ *n* The size of a bus, which determines how much data can be transmitted; for example, a 64-bit bus can transmit 64 bits of data.

byte /baɪt/ *n* A unit of computer information, consisting of a group of eight bits. See also **kilobyte**, **megabyte**, **gigabyte**, **terabyte**.

C

C /siː/ *n* A high-level programming language developed in 1972 at AT&T Bell Labs. It is used to write system software, graphics and commercial applications.

C# /ˌsiː ˈʃɑːp/ *n* A simplified version of C and C++, developed by Microsoft for applications on the Web.

C++ /ˌsiː ˈplʌs ˈplʌs/ *n* An object-oriented version of C, widely used to develop enterprise and commercial applications. The programmer gives each object (e.g. a piece of text, a graphic or a table) functions which can be altered without changing the entire program.

cable modem /ˈkeɪbəl ˌməʊdem/ *n* A modem designed to operate through a cable TV line.

call centre /ˈkɔːl ˌsentə / *n* A large office in which a company's employees provide information to its customers, or sell or advertise its goods or services by telephone.

carbon copy (Cc) /ˈkɑːbən ˌkɒpi/ *n* Addresses on the Cc: line of an email program will receive the same message, and the recipients will be able to see the identity of the other recipients.

Cascading Style Sheets (CSS) /kæsˈkeɪdɪŋ ˌstaɪl ˌʃiːts/ n A mechanism for adding style (e.g. fonts, colours, spacing) to web documents.

cathode ray tube (CRT) /ˈkæθəʊd ˌreɪ ˌtjuːb/ *n* The picture tube of old PC monitors, made of glass and containing a vacuum. In a colour monitor, the screen surface is coated with triads of red, green and blue phosphor. Three electron beams energize the phosphor dots, causing them to emit coloured light from which the picture is formed.

CD ripper /ˌsiːdiː ˈrɪpə/ *n* A program that extracts music tracks and saves them on disk.

CD-R /ˌsiːdiː ˈɑː/ *n* A write-once CD which lets you duplicate music and data CDs. Short for *compact disc recordable*.

CD-ROM /ˌsiːdiː ˈrɒm/ *n* A 'read-only' CD, meaning you cannot change data stored on it. Short for *compact disc read-only memory*.

CD-RW /ˌsiːdiː ˌɑː ˈdʌbəljuː/ *n* A CD that allows audio or data to be written, read, erased, and rewritten. Short for *compact disc re-writable*.

cell /sel/ *n* An intersection of a column and a row in a spreadsheet.

cell phone /ˈsel ˌfəʊn/ *n* American term for *mobile phone*. The term cell comes from the fact that the phone calls are made through base stations (antennae) which divide the coverage area into cells. As you move from cell to cell, the calls are transferred to different base stations; this is called *roaming*.

central processing unit (CPU) /ˌsentrəl ˈprəʊsesɪŋ ˌjuːnɪt/ *n* The processor chip that performs the basic operations of a computer; like the 'brain' of the computer. Its basic components are the control unit, the arithmetic logic unit and the registers.

character /ˈkærəktə/ *n* A symbol available on the keyboard (letter, number or blank space).

chat /tʃæt/ *n* A real-time interactive conversation on the Internet.

chat room /ˈtʃæt ˌruːm/ *n* A channel where users can communicate with each other in real time.

chip /tʃɪp/ *n* A tiny piece of silicon containing complex electronic circuits. Chips are used to make the hardware components of a computer.

Chip and Pin /ˌtʃɪp ˌænd ˈpɪn/ *n* A secure method of paying with credit cards. Instead of using a signature to verify payments, customers are asked to enter a four-digit PIN (personal identification number).

click /klɪk/ *v* To press and release the left button on a mouse.

client program /ˈklaɪənt ˌprəʊgræm/ *n* Software running on your PC, used to connect and obtain data from a server.

client-server /ˈklaɪənt,sɜːvə/ *n* A network architecture in which various client programs all connect to a central server to obtain information or to communicate.

clip art /ˈklɪp ˌɑːt/ *n* Ready-made pictures.

clipboard /ˈklɪpbɔːd/ *n* A holding place for text or graphics that you have just cut or copied.

COBOL /ˈkəʊbɒl/ *n* A high-level programming language developed in 1959 and mainly used for business applications. Short for *Common Business-Oriented Language*.

coding /ˈkəʊdɪŋ/ *n* The process of writing instructions for a computer.

colour depth /ˈkʌlə ˌdepθ/ *n* The number of bits used to hold a colour pixel; this determines the number of colours that a monitor can display.

colour palette /ˈkʌlə ˌpælət/ *n* The collection of colours available in a system.

colour picker /ˈkʌlə ˌpɪkə/ *n* A tool used to select a specific colour in a photo; also called an *eyedropper*.

column /ˈkʌləm/ *n* A vertical line of boxes labelled with a letter in a spreadsheet program.

command /kəˈmɑːnd/ *n* An instruction for a computer.

compact disc (CD) /kəmˌpækt ˈdɪsk/ *n* A storage device which uses optical laser technology. Its storage capacity is from 650MB to 700MB.

compatible /kəmˈpætɪbəl/ *adj* The ability of a device or program to work with another device or program. Two PCs are compatible if they can run the same software. Programs are compatible if they use the same data formats.

compiler /kəmˈpaɪlə/ *n* A special program that converts a source program (written in a high-level language) into object code (machine code) in one go.

compositing /kəmˈpɒzɪtɪŋ/ *n* Combining parts of different images to create a single image.

compression /kəmˈpreʃən/ *n* The process which makes computer data smaller so the information takes up less space and may be transmitted in less time. Compressed files have extensions like .zip, .arj, and .sit.

computer security specialist /kəmˌpjuːtə sɪˈkjʊərəti ˌspeʃəlɪst/ *n* Someone who works with companies to build secure computer systems.

computer-aided design (CAD) /kəmˌpjuːtə ˌeɪdɪd dɪˈzaɪn/ *n* Software, and sometimes special-purpose hardware, used by engineers and architects to design everything from cars and planes to buildings and furniture.

computer-to-plate (CTP) /kəmˌpjuːtə tə ˈpleɪt/ *n* An imaging technology used in modern commercial printing, in which DTP files are output directly to the printing plates without using film as an intermediate step.

configuration /kənˌfɪgəˈreɪʃən/ *n* The components of a computer system.

configure /kənˈfɪgə/ *v* To set up a computer device or a program to be used in a particular way.

console game /ˈkɒnsəʊl ˌgeɪm/ *n* A game played on a video game console, such as PlayStation or Xbox 360, and displayed on a television or similar audio-video system.

control unit (CU) /kənˈtrəʊl ˌjuːnɪt/ *n* A component of the CPU which coordinates all the other parts of the computer system. This unit is also responsible for fetching instructions from the main memory and determining their type.

cookies /ˈkʊkiz/ *n* Small files used by web servers to know if you have visited their site before. Cookies can store user information but do not read your hard disk.

cracker /ˈkrækə/ *n* An intruder who breaks into computer systems for fun, to steal information, or to propagate viruses. Compare with **hacker**.

crash /kræʃ/ **1** *n* A serious failure which usually requires operator attention before the computer system can be restarted. **2** *v* When a hard disk fails, it is said to have crashed.

crop /krɒp/ *v* To cut down the dimensions of a picture.

cursor control keys /ˌkɜːsə kənˈtrəʊl ˌkiːz/ *n* They include arrow keys that move the insertion point up, down, right and left, and keys such as *End*, *Home*, *Page Up* and *Page Down*, used within a word processor to move around a long document.

CU-SeeMe /ˌsiː ˌjuː ˌsiː ˈmiː / *n* A video-conferencing program from Cornell university.

cybercafé /ˈsaɪbəˌkæfeɪ/ *n* A place where you can use computers with internet access for a fee; also called an *internet café*.

cybercrime /ˈsaɪbəˌkraɪm/ *n* Crimes perpetrated over the Net.

cyberculture /ˈsaɪbəˌkʌlʃə/ *n* Culture emerging from the use of ICT systems.

cyberslacker /ˈsaɪbəˌslækə / *n* An employee who uses his company's internet connection during working hours to chat with friends, play games, etc.

cyberspace /ˈsaɪbəˌspeɪs/ *n* A term originated by William Gibson in his novel *Neuromancer*, now used to refer to the virtual world of computers and the Internet.

cyberstalking /ˈsaɪbəˌstɔːkɪŋ/ *n* Online harassment or abuse, mainly in chat rooms and forums.

cyborg /ˈsaɪbɔːg/ *n* A term invented by M. Clynes and N. Kline in 1960 to describe a cybernetic organism, a being that is part robot, part human.

D

data /ˈdeɪtə/ *n* Information in an electronic form that can be stored and processed by a computer.

data processing /ˈdeɪtə ˌprəʊsesɪŋ/ *n* The performing of operations on data to obtain information or solutions to a problem.

data transfer rate /ˌdeɪtə ˈtrænsfɜː ˌreɪt/ *n* The average speed at which data can be transmitted from one device to another, often measured in megabytes per second.

database /ˈdeɪtəbeɪs/ *n* A file of structured data.

database program /ˈdeɪtəbeɪs ˌprəʊgræm/ *n* An applications program used to store, organize and retrieve a large collection of data. Among other facilities, data can be searched, sorted and updated.

debug /ˌdiːˈbʌɡ/ *v* To correct program errors, or *bugs*.

debugger /ˌdiːˈbʌɡə/ *n* A program used to test and debug other programs.

debugging /ˌdiːˈbʌɡɪŋ/ *n* The techniques of detecting and correcting errors (or *bugs*) which may occur in programs.

decryption /dɪˈkrɪpʃən/ *n* The process of decoding (deciphering) secret data.

dedicated keys /ˈdedɪkeɪtɪd ˌkiːz/ *n* Special keys used to issue commands or to produce alternative characters, e.g. the *Ctrl* key or the *Alt* key.

desk accessory /ˌdesk əkˈsesəri/ *n* A mini application available on the Apple Menu, e.g. a calculator. In Palm OS, it is a program that you can launch from any program without having to exit the running program.

desktop PC /ˈdesktɒp piːˌsiː/ *n* A computer designed to be placed on a desk, used as a home computer or as a workstation for group work.

desktop publishing (DTP) /ˌdesktɒp ˈpʌblɪʃɪŋ/ *n* The use of a computer system for all steps of document production, including typing, editing, graphics and printing.

device driver /dɪˈvaɪs ˌdraɪvə/ *n* A program that allows a hardware device, such as a printer, to communicate with a computer.

dial-up connection /ˈdaɪəlʌp kəˌnekʃən/ *n* A form of internet access through which the client uses a modem connected to a computer and a telephone line to dial into an Internet Service Provider. A dial-up connection is slower than a broadband connection like ADSL.

dialog box /ˈdaɪəlɒɡ ˌbɒks/ *n* A message box requiring information from the user.

digital /ˈdɪdʒɪtəl/ *adj* Describes a system that performs operations by means of digits, represented as binary numbers (1s and 0s). The opposite of digital is *analogue*.

digital camera /ˌdɪdʒɪtəl ˈkæmərə/ *n* A still camera that records images in digital form. Instead of using the film found in a traditional camera, it uses a flash memory card.

digital certificate /ˌdɪdʒɪtəl səˈtɪfɪkət/ *n* A file that identifies a user or a web server; like a digital identification card.

digital radio /ˌdɪdʒɪtəl ˈreɪdiəʊ/ *n* Radio technologies which carry information as digital signals; also known as *digital audio broadcasting* (DAB).

digital TV /ˌdɪdʒɪtəl tiːˈviː/ *n* A way of transmitting pictures by means of digital signals, in contrast to the analogue signals used by traditional TV.

digital video camera /ˌdɪdʒɪtəl ˈvɪdiəʊ ˌkæmərə/ *n* A camera that records moving images and converts them into digital data; also called a *camcorder*.

digitize /ˈdɪdʒɪtaɪz/ *v* To translate into digital form, i.e. convert information into binary codes (1s and 0s) so that it can be processed by a computer. It is possible to digitize images, sound and video.

directory /dɪˈrektəri/ *n* An alphabetical or chronological list of files on a disk. Also known as *catalogue*.

disk drive /ˈdɪsk ˌdraɪv/ *n* A device that reads and writes data on disks. Magnetic drives read magnetic disks (e.g. hard disks), and optical drives use a laser beam to read optical discs (e.g. CDs and DVDs).

disk partitioning /ˈdɪsk pɑːˌtɪʃənɪŋ/ *n* The process of dividing a hard disk into isolated sections. In Windows, each partition will behave like a separate disk drive. This is particularly useful if you want to install more than one operating system (e.g. you can have one partition for Windows and another for Linux).

display /dɪˈspleɪ/ **1** *n* A screen or monitor. **2** *v* To show text and graphics on a screen.

DivX /ˈdɪvex/ *n* A format used to compress and distribute movies on DVD or over the Net.

Dock /dɒk/ *n* A set of icons at the bottom of the Macintosh screen that give you instant access to the things you use most.

domain name /dəˈmeɪn ˌneɪm/ *n* A name that identifies internet sites, consisting of two or more parts separated by dots. For example, in the web address http://www.ibm.com, the part on the left (.ibm) is the most specific (a subdomain); the part on the right (.com) is the most general (a primary domain); this can be a country (e.g. .fr for France, .uk for United Kingdom, .es for Spain), or the type of organization (e.g. .com for commercial, .org for organization, .edu for educational, .net for network, or gov for government). An IP address (e.g. 194.179.73.2) is translated into a domain name by a Domain Name System.

dot-matrix /ˌdɒt ˈmeɪtrɪks/ *n* A regular pattern of dots.

dot-matrix printer /ˌdɒt ˌmeɪtrɪks ˈprɪntə/ *n* A printer that uses pins to print an array of dots; used to print multi-part forms, self-copying paper and continuous-form labels.

double click /ˈdʌbəl ˌklɪk/ *v* To press and release the left button on a mouse twice, in rapid succession.

download /ˌdaʊnˈləʊd/ *v* To copy files from a host computer to your own computer; compare with **upload**.

drag /dræɡ/ *v* To select a block of text or an object with the mouse button and then move the mouse while keeping the button pressed down.

Dreamweaver /ˌdriːmˈwiːvə/ *n* A program from Adobe (originally created by Macromedia), used for building websites.

drop-down menu /ˌdrɒp ˌdaʊn ˈmenjuː/ *n* A list of options that appears below a menu item when selected; also called a *pull-down menu*.

DTP operator /diːtiːˈpiː ˌɒpəreɪtə/ *n* Someone who uses page-layout software to prepare electronic files for publication.

dual-core processor /ˌdjuːəl ˌkɔː ˈprəʊsesə/ *n* A CPU that combines two execution cores (processors) onto a single chip

DVD /ˌdiːviːˈdiː/ *n* A Digital Versatile (or Video) Disc that uses optical technology to store large amounts of audio-visual material. Whereas CDs use only one side, DVDs can be recorded on both sides as well as in dual layers. A basic DVD can hold 4.7GB, and a dual layer DVD can hold 17GB.

DVD burner /ˌdiːviːdiː ˈbɜːnə/ *n* A DVD drive that records information by burning via a laser to a blank DVD disc.

DVD-R /ˌdiːviːdiː ˈɑː/ *n* A recordable (write-once) DVD, for both movies and data.

DVD-ROM /ˌdiːviːdiː ˈrɒm/ *n* A read-only DVD disc used in DVD computer drives for data archival as well as interactive content (e.g. an encyclopedia, a movie, etc.).

DVD-RW /ˌdiːviːdiː ˌɑː ˈdʌbljuː/ *n* A re-writable (write-many) DVD, for movies and data.

E

eBay /ˈiːbeɪ/ *n* An online auction and shopping website where you can buy and sell things.

e-book /ˈiːbʊk/ *n* The electronic counterpart of a printed book.

e-card /ˈiːkɑːd/ *n* A digital greeting card.

e-cash /ˈiːkæʃ/ *n* Money available as an electronic account, used in internet commerce.

e-commerce /ˌiːˈkɒmɜːs/ *n* The buying and selling of products on the Internet

edit /ˈedɪt/ *v* To make changes and corrections to text and graphics. Well-known editing techniques are: select, undo, copy, cut, and paste.

e-learning /ˌiːˈlɜːnɪŋ/ *n* Instruction via computers.

email /ˈiːmeɪl/ *n* **1** *n* A facility which allows users to exchange messages electronically; short for *electronic mail*. **2** *v* To send a message by email.

email address /ˈiːmeɪl əˌdres/ *n* A unique address used to receive and send email. This is a typical format: jmartin1984@telefonica.net, where 'jmartin1984' is the user name, @ means 'at', 'telefonica' is the Internet Service Provider, and 'net' means the server is a network provider.

email client /ˈiːmeɪl ˌklaɪənt/ *n* A program used to read and send email from a computer.

embedded /ɪmˈbedɪd/ *adj* Inserted into; fixed into the surface of something.

emoticon /ɪˈməʊtɪkɒn/ *n* See **smiley**.

encrypt /ɪnˈkrɪpt/ *v* To encode data so that unauthorized users can't read it.

encryption /ɪnˈkrɪpʃən/ *n* The process of saving and transmitting data in encoded form. Data encryption and passwords are important for network security, particularly when sending confidential information such as credit card numbers.

e-pal /ˈiːpæl/ *n* A friend you write email to.

eraser /ɪˈreɪzə/ *n* A tool used to delete the part of the picture you drag it over.

ergonomics /ˌɜːɡəˈnɒmɪks/ *n* The study of how people interact safely and efficiently with machines and their work conditions.

e-signature /ˌiːˈsɪɡnətʃə/ *n* The electronic equivalent of a hand-written signature.

e-tailer /ˈiːteɪlə/ *n* An electronic retailer, or online store.

Ethernet /ˈiːθənet/ *n* A method of connecting computers in a LAN. Fast Ethernet can send data at 100 megabits per second. Most computers come with Ethernet ports that connect internally to circuits on the motherboard.

Excel /ɪkˈsel/ *n* A spreadsheet program from Microsoft.

execute /ˈeksɪkjuːt/ *v* To perform an action, as in executing a program or a command; the same as *run*.

expandable /ɪkˈspændəbəl/ *adj* Upgradeable; able to increase in size. For example, RAM is expandable in most computers, which means you can add extra chips, usually contained in small circuit boards called *dual in-line memory modules*, or *DIMMs*.

expansion card /ɪkˈspænʃən ˌkɑːd/ *n* A printed circuit board that can be inserted into an expansion slot to add features like sound, memory, and network capabilities; the same as *expansion board*.

expansion slots /ɪk'spænʃən ˌslɒts/ *n* The connectors that allow the user to install expansion cards to improve the computer's performance.

eyegaze system /'aɪgeɪz ˌsɪstəm/ *n* A system activated by the user's eye movements.

e-zine /'iːziːn/ *n* An electronic magazine.

F

FAQ /ˌefeɪ'kjuː/ *n* Frequently Asked Questions, a file or web page containing answers to questions asked by internet users or visitors to a website.

fax /fæks/ *n* A facsimile machine that operates by scanning a paper document so that the image is sent to a receiving machine which produces a copy of the original.

fibre optic communication /ˌfaɪbə ˌɒptɪk kəˌmjuːnɪ'keɪʃən/ *n* A way of transmitting information at high-speed by sending light through an optical fibre (made of glass or plastic). Fibre optic cables are used to transmit internet, cable TV and phone signals.

field /fiːld/ *n* A unit of information in a record. In a database, information is entered via fields.

file /faɪl/ *n* **1** A collection of records in a database. **2** A section of information stored on disk – a document or a program.

file server /'faɪl ˌsɜːvə/ *n* A fast computer that stores the programs and data files shared by users in a network.

File Transfer Protocol (FTP) /ˌfaɪl ˌtrænsfɜː 'prəʊtəkɒl/ *n* A standard for transferring files from one computer to another over a network.

filter /'fɪltə/ *n* A special effect that can be applied to pictures.

filtering program /'fɪltərɪŋ ˌprəʊgræm/ *n* Software designed to restrict the access to specific aspects of the Web.

Find and Replace /ˌfaɪnd ənd rɪ'pleɪs/ *n* A command that lets you find a word or phrase in a document and change it to new text.

Firefox /'faɪəfɒks/ *n* A web browser, part of the open-source Mozilla project.

firewall /'faɪəwɔːl/ *n* A software and/or hardware device that allows limited access to an internal network from the Net. This prevents intruders from stealing or destroying confidential data.

firmware /'fɜːmweə/ *n* Permanent software instructions contained in the ROM.

flame /fleɪm/ *n* An angry or insulting comment on a discussion group.

Flash /flæʃ/ *n* **1** The Adobe Flash Player. **2** The Adobe Flash Professional multimedia authoring program, used to create animations and advertisements. It supports a scripting language called ActionScript, and the streaming of audio and video.

flash card reader /'flæʃ ˌkɑːd ˌriːdə/ *n* A device that reads and writes a flash memory card.

flash drive /'flæʃ ˌdraɪv/ *n* A USB storage device, small enough to fit on a key ring, used to store and transport computer data.

flash memory /ˌflæʃ 'meməri/ *n* A type of non-volatile memory that can be erased and reprogrammed.

flatbed scanner /'flætbed ˌskænə/ *n* A scanner with a glass scanning surface on which objects are placed; similar to a photocopier.

flat-rate internet /ˌflætreɪt 'ɪntənet/ *n* Access to the Internet at any time of the day, at a fixed and cheap tariff.

Flickr /'flɪkə/ *n* A website where users can share photos.

floppy disk /'flɒpi ˌdɪsk/ *n* A disk made of a flexible plastic material upon which data is stored on magnetic tracks. Also known as a *diskette*. A floppy disk drive uses 3.5" disks.

flowchart /'fləʊtʃɑːt/ *n* A diagram which shows the logical steps of a computer program.

folder /'fəʊldə/ *n* A directory that holds programs, data files and other folders.

font /fɒnt/ *n* The shape, style and size of a particular typeface, e.g. **Times Bold at 10pt**.

footer /'fʊtə/ *n* Customized text printed in the bottom margin of a document.

format /'fɔːmæt/ **1** *n* The layout of a document, including page numbers, line spaces, margins, paragraph alignment, headers and footers, etc. **2** **format a disk** *v* To prepare a disk for use. When a disk is initialized, the operating system marks tracks and sectors on its surface.

formatting toolbar /'fɔːmætɪŋ ˌtuːlbɑː/ *n* A toolbar with icons that allow you to edit and style your text. For example, you can change font, align text, increase or decrease indentation, etc.

formula /'fɔːmjələ/ *n* A mathematical equation that helps you calculate and analyse data.

FORTRAN /'fɔːtræn/ *n* The first high-level programming language and compiler, developed in 1954 by IBM. Today, it is still used in mathematics, science, and engineering. Short for *FORmula TRANslation*.

fractals /'fræktəlz/ *n* Geometrical patterns that are repeated at small scales to generate irregular shapes, some of which describe objects from nature.

fragmentation /ˌfrægmən'teɪʃən/ *n* The condition of a hard disk in which files are divided into pieces scattered around the disk. This occurs naturally after creating, deleting and modifying many files. When the operating system cannot find enough contiguous space to store a complete file, the file is divided into several separated fragments. As disk fragmentation increases, disk efficiency starts decreasing.

frames /freɪmz/ *n* **1** Rectangular areas that allow the display of different pages in the same browser window. **2** Single pictures in films.

Freehand /'friːhænd/ *n* A Macromedia program for creating vector graphics, which use geometrical primitives such as points, lines, curves and polygons to represent images.

freeware /'friːweə/ *n* Software that is available free of charge, but protected by copyright.

FrontPage /ˌfrʌnt'peɪdʒ / *n* A web editor from Microsoft, used for designing web pages.

function /'fʌŋkʃən/ *n* A ready-to-use formula that helps you perform a specialized calculation, e.g. SUM, AVERAGE, etc.

function keys /'fʌŋkʃən ˌkiːz/ *n* Keys that appear at the top of the keyboard and can be programmed to do special tasks.

G

gadget /'gædʒɪt/ *n* A small hardware device. Synonymous with *gizmo* (slang).

game controller /'geɪm kənˌtrəʊlə/ *n* A device used to control video games.

game genre /ˌgeɪm 'ʒɑ̃ːnrə/ *n* A specific type or category of game. For example, a game in which the player solves puzzles would fall into the Puzzle game genre. Other genres are: Action, Adventure, Fighting, First-person shooter, Role-playing, Simulation, Sports, Strategy, etc.

game platform /ˌgeɪm 'plætfɔːm/ *n* An electronic device on which video games are played. Examples are personal computers and game consoles.

Geographic Information System (GIS) /dʒiːəˌgræfɪk ɪnfə'meɪʃən ˌsɪstəm/ *n* A type of graphics software that allows us to analyse geographic data and then make maps, plan the use of land, predict natural disasters, etc.

gigabyte /'gɪgəbaɪt/ *n* 1,024 megabytes.

gigahertz /'gɪgəhɜːts/ *n* A unit of one thousand million hertz, or cycles per second, used to measure processor speed.

Global Positioning System (GPS) /ˌgləʊbəl pə'zɪʃənɪŋ ˌsɪstəm/ *n* A navigation system formed by various satellites orbiting the earth and their corresponding receivers on the earth. It allows GPS receivers to determine their location, speed and direction.

Google /'guːgəl/ **1** *n* A popular search engine on the Web. **2** *v* To search the Web for something.

grammar checker /'græmə ˌtʃekə/ n A software utility that analyses the grammar of a written text.

graphical user interface (GUI) /ˌgræfɪkəl 'juːzə ˌɪntəfeɪs/ *n* A user-friendly interface based on graphics. A GUI uses a WIMP environment: windows, icons, menus and pointer. Typical examples are the Mac OS and Microsoft Windows.

graphics tablet /'græfɪks ˌtæblət/ n An input device which allows the user to enter drawings and sketches into a computer

H

hacker /'hækə/ *n* Someone who invades a network's privacy. Originally, all skilled programmers were known as hackers, but in the 1990s, the term became synonymous with *cracker*, a person who breaks security on computers. Today, the general public uses hacker for both. In the computer industry, hackers are known as *white hats* and crackers as *black hats* or *darkside* hackers.

handheld game /'hændheld ˌgeɪm/ *n* A game played on portable gaming devices, such as the Sony PSP and the Nintendo DS.

handheld scanner /'hændheld ˌskænə/ *n* A scanner that is moved by hand, ideal for capturing small pictures, logos and bar codes.

hard disk /'hɑːd ˌdɪsk/ *n* See **hard drive**.

hard drive /'hɑːd ˌdraɪv/ *n* A magnetic storage device that reads and writes data on metal disks (called platters) inside a sealed case. A hard drive is commonly known as a hard disk. Strictly speaking, drive refers to the entire unit, containing multiple platters, a read/write head and a motor, while hard disk refers to the storage medium itself.

hardware /'hɑːdweə/ *n* The physical units which make up a computer system. See **software**.

hardware engineer /'hɑːdweə endʒɪˌnɪə/ *n* Someone who designs and develops IT devices.

header /'hedə/ n Customized text printed in the top margin of a document.

help desk technician /'help ˌdesk tekˌnɪʃən/ *n* Someone who helps end users with their computer problems in person, by email or over the phone.

hertz /hɜːts/ *n* A unit of frequency equal to one cycle per second, named after Heinrich Hertz.

high-level language /ˌhaɪ ˌlevəl 'læŋgwɪdʒ/ *n* A language in which each statement represents several machine code instructions, e.g. COBOL, Pascal or C.

home cinema /ˌhəʊm 'sɪnəmə/ *n* A system that tries to reproduce the cinema experience in the home. It is also called *home theatre* and typically includes a large-screen TV, a hi-fi system with speakers for surround sound, and a DVD recorder.

home page /'həʊm ˌpeɪdʒ/ *n* **1** The first page on a website, that usually contains links to other pages. **2** The default start-up page on which a web browser starts.

host /həʊst/ *n* A computer containing data or programs that other computers can access via a network or modem.

hotspot /'hɒtspɒt/ n The geographic boundary covered by a Wi-Fi wireless access point.

HTML /ˌeɪtʃtiːem'el/ *n* The language used to create hypertext documents (e.g. web pages); short for *Hypertext Markup Language*.

HTML tags /ˌeɪtʃtiːemel 'tægz/ *n* The codes used to define text fonts, format paragraphs, add links, etc. HTML tags are surrounded by the angle brackets < and >.

HTTP /ˌeɪtʃtiːtiː'piː/ *n* The method by which web pages are transferred from a website to your PC; http appears at the beginning of web addresses and means *hypertext transfer protocol*.

hybrid hard disk /ˌhaɪbrɪd 'hɑːd ˌdɪsk/ *n* A hard disk with integrated flash memory, intended for new laptops and mobile PCs.

hyperlink /'haɪpəlɪŋk/ *n* A text, image or button that, when clicked, takes you to other destinations on the Web.

hypermedia /'haɪpəmiːdiə/ *n* A form of enriched multimedia which supports linking graphics, sound, and video elements in addition to text elements.

hypertext /'haɪpətekst/ *n* Text that contains links to other documents.

I

icon /'aɪkɒn/ *n* A picture representing an object, such as a document, program, folder or hard disk.

ICT system /ˌaɪsiːtiː 'sɪstəm/ *n* A system that uses information and communications technologies.

IM server /ˌaɪem 'sɜːvə/ *n* A central system that provides presence information about online users, and passes instant messages between them.

iMac /'aɪmæk/ *n* A desktop computer from Apple, intended for home, school, and small offices.

imagesetter /'ɪmɪdʒˌsetə/ *n* A professional printer that generates high-resolution output on paper or microfilm.

inch /ɪnʃ/ *n* The equivalent of 2.54 cm, or 72.27 points. It is represented by the symbol ".

indentation /ˌɪnden'teɪʃən/ *n* The space between the page margins and where the text aligns.

InDesign /'ɪndɪzaɪn/ *n* A desktop publishing program created by Adobe Systems.

ink cartridge /'ɪŋk ˌkɑːtrɪdʒ/ *n* A replaceable container that holds the ink of an inkjet printer.

inkjet printer /'ɪŋkdʒet ˌprɪntə/ *n* A printer that generates an image by spraying tiny drops of ink at the paper. By heating the ink within the print head, individual drops are expelled to make a matrix of dots on the paper.

input /'ɪnpʊt/ **1** *n* The process of transferring information into the memory from a peripheral unit. **2** *v* To transfer data, or program instructions, into the computer.

input devices /'ɪnpʊt dɪˌvaɪsɪz/ *n* Units of hardware which allow the user to enter information into the computer, e.g. the keyboard, mouse, voice recognition devices, etc.

Instant Messaging (IM) /ˌɪnstənt 'mesɪdʒɪŋ/ *n* Exchanging text messages in real-time between two or more people logged into IM services such as AIM, Windows Live Messenger and Yahoo! Messenger. Modern IM services also have audio and video capabilities.

Intel /'ɪntel / *n* The company that designs and produces the processors used in most PCs.

Intel Core 2 Duo /ˌɪntel ˌkɔː ˌtuː 'djuːəʊ/ *n* Technology that includes two cores, or processors, into a single chip, offering twice the speed of a traditional chip.

interactive whiteboard /ɪntəˌæktɪv 'waɪtbɔːd/ *n* A touch-sensitive projection screen that allows the user to control a computer directly, by touching the board instead of using a keyboard. Used in presentation situations such as teaching.

interface /'ɪntəfeɪs/ *n* Channels and control circuits which provide a connection between the CPU and the peripherals. See also **user interface**.

Internet /'ɪntənet/ *n* A global network of computer networks which offers services such as email, file transfer, online chats, newsgroups, and information retrieval on the Web. It evolved from the Arpanet of the 70s and uses the TCP/IP protocol.

internet auction /ˌɪntənet 'ɔːkʃən/ *n* A website on which bids are received and transmitted electronically.

Internet Explorer /ˌɪntənet ɪk'splɔːə/ *n* A popular web browser from Microsoft.

Internet Service Provider (ISP) /ˌɪntənet 'sɜː vɪs prəʊˌvaɪdə/ *n* The company which gives you access to the Internet.

internet telephony /ˌɪntənet tɪ'lefəni/ *n* See **VoIP**.

internet TV /ˌɪntənet tiː'viː/ *n* A TV set used as an internet device.

interpreter /ɪn'tɜːprɪtə/ *n* A special program that translates the source code line by line, as the program is running.

Intranet /'ɪntrənet/ *n* A company network that uses public internet software but makes the website only accessible to employees and authorized users.

invoice /'ɪnvɔɪs/ *n* A document showing the items purchased, quantities, prices, etc., and requesting payment for a credit order.

IP address /aɪ'piː əˌdres/ *n* A number which identifies a computer on the Internet. Every computer on the Net has a unique IP address, e.g. 194.179.73.2.

IP spoofing / aɪˌpiː 'spuːfɪŋ/ *n* Making one computer look like another to gain unauthorized access.

iPhone /'aɪfəʊn/ *n* A device from Apple that combines three products in one: an iPod, a mobile phone and an internet communicator.

iPod /'aɪpɒd/ *n* A family of portable media players from Apple. Popular models include the iPod Nano, the iPod Shuffle and the full-sized iPod that can also be used as a portable hard disk.

iTunes /'aɪtjuːnz/ *n* A program from Apple that lets you play and organize music and video files, on computer or on an iPod. With an internet connection, iTunes can also connect to the iTunes Store in order to download purchased music, videos and podcasts.

J

Java /'dʒɑːvə/ *n* The programming language from Sun Microsystems for building internet applications. Java programs (called applets) let you watch animated characters and moving text, play music, etc.

Java ME /ˌdʒɑːvə e'miː/ *n* The Java platform, Micro Edition, used to create applications that run on mobile phones, PDAs, TV set-top boxes, and printers. For example, many phones are configured to use Java games.

joystick /'dʒɔɪstɪk/ *n* An input device with a vertical lever, used in computer games.

JPEG /'dʒeɪpeg/ *n* A standard for compressing and decompressing image files; developed by the Joint Photographic Experts Group. A .jpg extension is added to many image files on the Web.

K

kerning /'kɜːnɪŋ/ *n* The process of adjusting the spaces between letters to achieve even, consistent letter spacing.

keyboard /'kiːbɔːd/ *n* A set of keys on a terminal or computer, including the standard typewriter keys (for letters and numbers), function keys and several special keys.

kilobit /'kɪləbɪt/ *n* One thousand bits.

kilobyte /'kɪləbaɪt/ *n* A unit for measuring the memory or disk space in thousands of bytes. Also called **k**. Equals 1,024 bytes.

L

laptop /'læptɒp/ *n* A small type of portable computer.

laser printer /'leɪzə ˌprɪntə/ *n* A printer that uses a laser beam to fix the ink (toner) to the paper.

lightpen /'laɪtpen/ *n* A highly sensitive photo-electric device which uses the screen as the positioning reference. The user can pass the pen over the surface of the screen to draw or modify images displayed on the screen.

link /lɪŋk/ *n* See **hyperlink**.

link up /ˌlɪŋk 'ʌp/ *v* To form a connection in order to operate together.

Linux /'lɪnəks/ *n* Open-source software developed under the GNU General Public License. This means anybody can copy its source code, change it and distribute it.

liquid crystal display /ˌlɪkwɪd 'krɪstəl dɪˌspleɪ/ *n* A flat-screen display made of two glass plates with a liquid crystal material between them. The crystals block the light in different quantities to create the image. Active-matrix LCDs use TFT (thin film transistor) technology, producing very sharp images.

lithium-ion battery /ˌlɪθiəm ˌaɪən 'bætəri/ *n* A type of a battery composed of Lithium, a metallic chemical element, used in PDAs, cameras and mobile phones.

load /ləʊd/ *v* To read program instructions into the main memory.

Local area network (LAN) /ˌləʊkəl ˌeəriə 'netwɜːk/ *n* A group of computer devices interconnected within a small physical area, like a home or office building.

log in/on /ˌlɒg 'ɪn/ /ˌlɒg 'ɒn/ *v* To gain access to a computer system or network.

log out/off /ˌlɒg 'aʊt/ /ˌlɒg 'ɒf/ *v* To sign off; to end a computer session. The opposite of *log in/on*.

login/logon /'lɒgɪn/ /'lɒgɒn/ *n* The process of identifying yourself when entering a computer system or network. You usually type your user name and password.

low-level language /ˌləʊ ˌlevəl 'læŋgwɪdʒ/ *n* A programming language that is very close to machine language. See **assembly language**.

M

Mac OS /ˌmæk əʊ'es/ *n* An operating system created by Apple and used on Macintosh computers.

MacBook /'mækbuk/ *n* A Macintosh notebook computer.

machine code /mə'ʃiːn ˌkəʊd/ *n* Binary code numbers; the only language that computers can understand directly.

magnetic storage devices /mæg'netɪk 'stɔːrɪdʒ dɪˌvaɪsɪz/ *n* devices that store data by magnetizing particles on a disk or tape (e.g. hard drive, tape drive).

magnetic tape /mæg'netɪk 'teɪp/ *n* A sequential storage device used for data collection, backup and archiving. A tape consists of a magnetic coating on a thin plastic strip.

mail merging /'meɪl ˌmɜːdʒɪŋ/ *n* The process of combining a database file with a word processor to personalize a standard letter.

mail server /'meɪl ˌsɜːvə/ *n* The computer where your Internet Service Provider stores your emails.

mailbox /'meɪlbɒks/ *n* The place where your email program stores new email for you.

mailing list /'meɪlɪŋ ˌlɪst/ *n* A system used to distribute email to many different subscribers at once.

main memory /ˌmeɪn 'meməri/ *n* The section which holds the instructions and data currently being processed; also referred to as the *immediate access store* or *internal memory*. PCs make use of two types of main memory: RAM and ROM.

Macintosh /'mækɪntɒʃ/ *n* A popular computer from Apple, introduced in 1984; the first computer with a graphical user interface.

mainframe /'meɪnfreɪm/ *n* The largest and most powerful type of computer. Mainframes process enormous amounts of data and are used in large installations.

malware /'mælweə/ *n* Malicious software, created to damage computer data. It includes viruses, worms, Trojan horses and spyware.

markup language /'mɑːkʌp ˌlæŋgwɪdʒ/ *n* A computer language that uses instructions, called markup tags, to format and link web documents.

marquee select tools /mɑːˌkiː sɪ'lekt ˌtuːlz/ *n* Tools used to select a particular part of an image.

master page /'mɑːstə ˌpeɪdʒ/ *n* A page you design which can be applied to any document page. You can place text and picture boxes, headers and footers, and page rules, etc. on a master page, which ensures a consistent look on all pages.

media player /'miːdiə ˌpleɪə/ *n* Software that plays audio, video or animation files.

megabit /'megəbɪt/ *n* A million binary digits (1,024 kilobits).

megabyte /'megəbaɪt/ *n* 1,024 kilobytes.

megahertz /'megəhɜːts/ *n* A unit of a million cycles per second, used to measure processor speed.

megapixel /'megəpɪksel/ *n* One million pixels.

memory card /'meməri ˌkɑːd/ *n* A removable module used to store images in digital cameras, to record voice and music on MP3 players, or to back up data on PDAs. They are made up of flash memory chips (e.g. CompactFlash, Secure Digital). See **flash memory**.

menu bar /'menjuː ˌbɑː/ *n* A row of words at the top of the screen that open up menus when selected.

message threads /'mesɪdʒ ˌθredz/ *n* A series of interrelated messages on a given topic.

microchip /'maɪkrətʃɪp/ *n* See **chip**.

Microsoft Access /ˌmaɪkrəsɒft 'ækses/ *n* A relational database management system.

Microsoft Office /ˌmaɪkrəsɒft 'ɒfɪs/ *n* An integrated package that includes some combination of Word, Excel, PowerPoint, Access and Outlook, along with various internet and other utilities.

MIDI /'mɪdi/ *n* A standard for connecting computers and musical instruments. MIDI files contain the .midi extension, short for *Musical Instrument Digital Interface*.

millisecond /'mɪlɪˌsekənd/ n One thousandth of a second.

mobile (phone) /'məʊbaɪl/ *n* (US: **cell phone**) A phone connected to the telephone system by radio, rather than by a wire.

modem /'məʊdem/ *n* A device that converts the digital signals used by computers into the analogue signals used by the telephone lines, thus allowing access to the Internet. Short for *MODulator/DEModulator*.

modem-router /'məʊdem ˌruːtə/ *n* A device that connects various computers (e.g. a home LAN) to the Internet.

monitor /'mɒnɪtə/ *n* An output device with a screen on which words or pictures can be shown. Also called a *display screen*.

motherboard /'mʌðəbɔːd/ *n* The main circuit board of a computer, which contains the processor, memory chips, expansion slots and controllers for peripherals, connected by buses.

mouse /maʊs/ *n* A small input device used to specify the position of the cursor or to make choices from menus. A *mechanical mouse* has a rubber or metal ball underneath that is rolled by the user. An *optical mouse* uses light (a laser) to detect the mouse's movement, and can be wired or wireless.

MP3 /ˌempiː'θriː/ *n* **1** A standard format that compresses music files, enabling them to be transmitted over the Net more easily. **2** A file containing a song or other audio data that is encoded using the MP3 standard.

MP3 player /empiː'θriː ˌpleɪə/ *n* A digital music player that supports the MP3 format.

MP4 player /empiː'fɔːr ˌpleɪə/ *n* A portable media player that plays video in the MPEG-4 format; it is like an MP3 player that can play video files.

MPEG /'empeg/ *n* A standard for compressing and decompressing video files; developed by the Moving Pictures Experts Group.

multi-format playback /mʌltiˌfɔːmæt 'pleɪbæk/ *n* The feature of a media player that makes it compatible with many file formats, including DVD-video, DivX, MP3 music or JPEG images.

multi-function printer /mʌltiˌfʌŋkʃən 'prɪntə/ *n* An 'all-in-one' device that can work as a printer, a scanner, a fax and a photocopier.

multimedia /ˌmʌlti'miːdiə/ *n* The integration of text, graphics, audio, video and animation in a single application.

multitasking /'mʌltiˌtɑːskɪŋ/ *n* The execution of several tasks at the same time.

multi-threaded /'mʌltiˌθredɪd/ *adj* Refers to a computer program that has multiple threads (parts), i.e. many different things processing independently and continuously. This enables the program to make the best use of available CPU power.

MySpace /'maɪspeɪs/ *n* A social networking site that allows users to share messages, interests, blogs, photos, music and videos with friends.

N

nanobot /'nænəʊˌbɒt/ *n* A microscopic robot, built by means of nanotechnology.

nanocomputer /ˌnænəʊkəm'pjuːtə/ *n* A molecule-sized computer, the size of a grain of sand, e.g. a quantum computer, a DNA computer, etc.

nanotechnology /ˌnænəʊtek'nɒlədʒi/ *n* The science of making small devices from single atoms and molecules.

nanotube /'nænəʊˌtjuːb/ *n* Extremely small tube made from pure carbon. Nanotubes are expected to be used in the development of materials for buildings, cars, airplanes, clothes, etc.

netiquette /'netɪket/ *n* 'Net etiquette'; good manners when communicating online.

NetMeeting /'netˌmiːtɪŋ/ *n* A VoIP and video-conferencing program from Microsoft.

Netscape Navigator /ˌnetskeɪp 'nævɪgeɪtə/ *n* A web browser developed by Netscape Communications.

network /'netwɜːk/ *n* A system of computer devices or 'nodes' (e.g. PCs and printers), interconnected so that information and resources can be shared by a large number of users.

network administrator /ˌnetwɜːk əd'mɪnɪstreɪtə/ *n* Someone who manages the hardware and software that comprise a network.

newsgroups /'njuːzgruːps/ *n* The public discussion areas which make up Usenet. The contents are contributed by people who send articles (messages) or respond to articles.

newsreader /'njuːzˌriːdə/ *n* A program that reads and sends articles to newsgroups.

nickname /'nɪkneɪm/ *n* A name used by a participant on mailing lists or chat sessions instead of the real name.

node /nəʊd/ *n* Any computer device in a network.

non-volatile memory /nɒnˌvɒlətaɪl 'meməri/ *n* Permanent memory, able to hold data without power. ROM and Flash memory are examples of non-volatile memory.

notebook computer /ˌnəʊtbʊk kəm'pjuːtə/ *n* A light, portable computer that is generally thinner than a laptop.

numeric keypad /njuːˌmerɪk 'kiːpæd/ *n* A small key section that appears to the right of the main keyboard and contains numeric and editing keys.

O

object-oriented programming /ˌɒbdʒɪkt ˌɔːrientɪd 'prəʊgræmɪŋ/ *n* A technique that allows the creation of objects that interact with each other and can be used as the foundation of others. Used to develop graphical user interfaces.

offline /ˌɒf'laɪn/ *adj or adv* Not connected to the Internet.

online /'ɒnlaɪn/ *adj or adv* Connected to the Internet.

online banking /ˌɒnlaɪn 'bæŋkɪŋ/ *n* Performing transactions and payments through a bank's website. Also known as *internet banking*.

onscreen keyboard /ɒnˌskriːn 'kiːbɔːd/ *n* A graphic representation of a keyboard on the computer screen, allowing people with mobility problems to type data using a joystick or pointing device.

open-source /'əʊpən ˌsɔːs/ *adj* Refers to the source code (of software) that is free and available to anyone who would like to use it or modify it.

operating system /'ɒpəreɪtɪŋ ˌsɪstəm/ *n* A set of programs that control the hardware and software of a computer system. Typical functions include handling input/output operations, running programs and organizing files on disks.

optical character recognition /ˌɒptɪkəl 'kærəktə rekəgˌnɪʃən/ *n* Technology that allows computers to recognize text input into a system with a scanner. After a page has been scanned, an OCR program identifies fonts, styles and graphic areas.

optical disc /ˌɒptɪkəl 'dɪsk/ *n* A storage device in which data is recorded as microscopic 'pits' by a laser beam. The data is read by photoelectric sensors which do not make active contact with the storage medium.

output /'aʊtpʊt/ **1** *n* The results produced by a computer. **2** *v* To transfer information from a CPU to an output device.

output devices /'aʊtpʊt dɪˌvaɪsɪz/ *n* The units of hardware which display the results produced by the computer (e.g. plotters, printers, monitors).

P

.pdf /ˌpiːdiː'ef/ *n* A portable document format from Adobe, commonly used to distribute text files over the Internet, and read with Acrobat Reader.

page description language /ˌpeɪdʒ dɪ'skrɪpʃən ˌlæŋgwɪdʒ/ *n* A computer language that describes how to print the text and images on each page of the document.

Page-layout program /ˌpeɪdʒ 'leɪaʊt ˌprəʊgræm/ *n* Application software used to import texts and illustrations, and to combine and arrange them all on a page; e.g. Adobe InDesign or QuarkXPress.

paint bucket /'peɪnt ˌbʌkɪt/ *n* A tool used to fill in an area with a colour.

Palm OS /ˌpɑːm əʊ'es/ *n* An operating system used on Palm hand-held devices.

palmtop /'pɑːmtɒp/ *n* A hand-held personal computer.

Pascal /pæs'kæl/ *n* A high-level language created in 1971, named after the mathematician Blaise Pascal. Its highly structured design facilitates the rapid location and correction of coding errors. Today, it's used in universities to teach the fundamentals of programming.

password /'pɑːswɜːd/ *n* A secret word which must be entered before access is given to a computer system or website.

paste /peɪst/ *v* To insert a copy of text or graphics, held in the computer's memory, at a chosen position of a document.

PC /ˌpiː'siː/ *n* A personal computer, which carries out processing on a single chip. PCs are often classified by size and portability: desktop PCs, laptops, tablet PCs and PDAs.

PC game /ˌpiːsiː 'geɪm/ *n* A game played on a personal computer.

peer-to-peer /ˌpɪə tə 'pɪə/ *n* A network architecture in which all the computers have the same capabilities, i.e. share files and peripherals, without requiring a separate server computer.

peer-to-peer file-sharing /ˌpɪə tə ˌpɪə 'faɪlˌʃeərɪŋ/ *n* A form of P2P networking which eliminates the need for central servers, allowing all computers to communicate and share resources (music files, videos, etc.) as equals.

pen drive /'pen ˌdraɪv/ *n* See **flash drive**.

peripherals /pə'rɪfərəlz/ *n* The units attached to the computer, classified into three types: input devices, output devices and storage devices.

Personal Digital Assistant (PDA) /ˌpɜːsənəl ˌdɪdʒɪtəl ə'sɪstənt/ *n* A tiny computer which can be held in one hand. The term PDA refers to a variety of hand-held devices, palmtops and pocket PCs. For input, you type at a small keyboard or use a stylus. It can be used as a personal organizer, a mobile phone or an internet device.

phishing /'fɪʃɪŋ/ *n* Getting passwords of online bank accounts or credit card numbers by using emails that look like real organizations, but are in fact fake; short for *password harvesting fishing*.

phosphor /'fɒsfə/ *n* The material or substance of the CRT screen that lights up when struck by an electron beam.

Photoshop /'fəʊtəʊʃɒp/ *n* An image manipulation program developed by Adobe Systems.

PictBridge /'pɪkbrɪdʒ/ *n* A technology developed by Canon that lets you print images from a memory card in a digital camera or a camera phone directly to the printer (no computer is necessary).

piracy /'paɪrəsi/ *n* The illegal copying and distribution of copyrighted programs and files.

pixel /'pɪksəl/ *n* The smallest unit on a display screen or bitmapped image (usually a coloured dot).

plasma screen /'plæzmə ˌskriːn/ *n* A display that generates images by a plasma discharge, which contains noble, non-harmful gases. It allows for larger screens and wide viewing angles.

platesetter /'pleɪtsetə/ *n* A machine that creates the printing plates.

platform-independent /ˌplætfɔːm ɪndɪ'pendənt/ *adj* Refers to software that can run on any computer system.

platter /'plætə/ *n* A magnetic plate, or disk, that constitutes part of a hard disk drive. There may be only one or several platters in a drive.

PlayStation /'pleɪˌsteɪʃən/ *n* A video game console from Sony.

plotter /'plɒtə/ *n* A graphics output device which is used to make various types of engineering drawings.

plug-ins /'plʌgɪnz/ *n* Special programs which extend the capabilities of a web browser so that it can handle audio, video, 3D and animation elements.

podcast /'pɒdkɑːst/ *n* An audio recording that is distributed by subscription (paid or unpaid) over the Internet using RSS feeds, for playback on mobile devices and PCs; coined from *iPOD* and *broadCAST*.

point /pɔɪnt/ *n* A unit used to measure font types and the distance between baselines. A point is a subdivision of a pica: there are 12 points in a pica and 72.27 points in an inch.

pointer /'pɔɪntə/ *n* **1** A small picture that follows the mouse movements. **2** The cursor which locates the insertion point on the screen, i.e. indicates where the next character will be displayed.

port /pɔːt/ *n* A socket or channel in the rear panel of the computer into which you can plug a wide range of peripherals: modems, scanners, digital cameras, etc. See **USB port**.

portable DVD player /ˌpɔːtəbəl diːviː'diː ˌpleɪə/ *n* A handheld device with a built-in DVD drive and a screen.

portable hard drive /ˌpɔːtəbəl 'hɑːd ˌdraɪv/ *n* An external hard drive that is connected to the USB or FireWire port of the computer.

portable media player /ˌpɔːtəbəl 'miːdiə ˌpleɪə/ *n* A handheld device that plays audio and video files.

PostScript /'pəʊsskrɪpt/ *n* A page description or graphics language developed by Adobe Systems. A PostScript font is any font defined in this language, e.g. Times or Helvetica.

power-line internet /ˌpaʊəlaɪn 'ɪntənet/ *n* A technology that provides low-cost internet access via the power plug.

PowerPoint /'paʊəpɔɪnt/ *n* A presentation graphics program from Microsoft.

Pretty Good Privacy (PGP) /ˌprɪti ˌgʊd 'prɪvəsi/ *n* A freeware program, written by Phil Zimmerman, designed to send email privately.

primary colours /ˌpraɪməri 'kʌləz/ *n* These are red, green and blue (RGB) in computers. Compare with the colours considered basic in inks (magenta, yellow and cyan).

primitives /'prɪmɪtɪvz/ *n* The basic shapes used to construct graphical objects: lines, polygons, etc.

print preview /ˌprɪnt 'priːvjuː/ *n* A function that shows how pages will look when printed.

printer /'prɪntə/ *n* An output device which converts data into printed form. The output from a printer is referred to as a *print-out* or *hard copy*.

printer driver /'prɪntə ˌdraɪvə/ *n* A program installed to control a particular type of printer.

printing plate /ˌprɪntɪn 'pleɪt/ *n* A metal surface that carries the image to be printed.

processor /'prəʊsesə/ *n* The chip that processes the instructions provided by the software. See **central processing unit** (**CPU**).

program /ˈprəʊɡræm/ n A set of instructions that tells the computer how to do a specific task. The task can be anything from the solution to a Maths problem to the production of a graphics package.

programmer /ˈprəʊɡræmə/ *n* Someone who writes computer programs.

programming /ˈprəʊɡræmɪŋ/ *n* The process of writing a program using a computer language.

protocol /ˈprəʊtəkɒl/ *n* A set of rules which determine the formats by which information may be exchanged between different systems.

proxy /ˈprɒksi/ *n* A computer server which controls the traffic between the Internet and a private network.

Q

QuarkXpress /ˌkwɑːk ɪkˈspres/ *n* A page layout application produced by Quark.

query /ˈkwɪəri/ *n* A request for data; in a database, a function that allows you to extract data according to certain conditions or criteria.

QuickTime /ˈkwɪktaɪm/ *n* Software from Apple that enables users to play, edit, and manipulate multimedia files.

R

radio tags /ˈreɪdiəʊ ˌtæɡz/ *n* Microchips attached to, or embedded into, products, animals or people, for the purpose of identification.

radio-frequency identification (RFID) /ˌreɪdiəʊ ˌfriːkwəntsi aɪdentɪfɪˈkeɪʃən/ *n* Technology that uses radio waves and chip-equipped tags (called RFID tags) to automatically identify people or things.

random access memory (RAM) /ˌrændəm ˈækses ˌmeməri/ *n* The part of the main memory which stores information temporarily while you are working. RAM requires a continuous power supply to retain information. Compare with **ROM**.

raster graphics /ˈrɑːstə ˌɡræfɪks/ *n* Images stored and displayed as pixels, which can become distorted when manipulated. Also called *bit-mapped graphics*.

read-only memory (ROM) /ˌriːdˈəʊnli ˌmeməri/ *n* Chips of memory containing information which is present and permanent. Also known as *firmware*.

read/write head /ˌriːd ˌraɪt ˈhed/ *n* The part of a disk drive that reads and writes data on a magnetic disk.

RealPlayer /ˈrɪəlpleɪə/ *n* A media player, created by RealNetworks, that plays a variety of audio and video formats.

real-time /ˈrɪəltaɪm/ *adj* Refers to something live, simultaneous (without delay), e.g. real-time chat.

reboot /riːˈbuːt/ *v* To restart the computer.

record /ˈrekɔːd/ *n* A unit of a file consisting of a number of interrelated data elements (fields).

register /ˈredʒɪstə/ *n* The component in the processor or other chip which holds the instruction from the memory while it is being executed.

relational database /rɪˈleɪʃənəl ˌdeɪtəbeɪs/ *n* A database system that maintains separate, related files (tables), but combines data elements from the files for queries and reports.

rendering /ˈrendərɪŋ/ *n* A technique that generates realistic reflections, shadows and highlights.

resolution /ˌrezəlˈuːʃən/ *n* The maximum number of pixels in the horizontal and vertical directions of the screen; also refers to the number of pixels per inch.

rewritable /riːˈraɪtəbl/ *adj* Able to be rewritten many times.

right click /ˌraɪt ˈklɪk/ *v* To press and release the right button on a mouse; this action displays a list of commands.

RIM /rɪm/ n An operating system used on BlackBerry communication devices, developed by Research In Motion.

ring topology /ˌrɪŋ tɒˈpɒlədʒi/ *n* One of the three principal topologies for a LAN, in which all devices are interconnected in a continuous loop, or ring.

ringtone /ˈrɪŋtəʊn/ *n* A digital sound file played by a telephone to announce an incoming call.

ripping /ˈrɪpɪŋ/ *n* Converting music tracks from a CD to the MP3 format.

rotation /rəˈteɪʃən/ *n* Turning an object around its axis.

router /ˈruːtə/ *n* A device used to transmit data between two computers or networks. See also **modem-router** and **wireless router**.

routine /ruːˈtiːn/ n A piece of code which performs a specific task in the operation of a program or system.

row /rəʊ/ *n* A horizontal line of boxes, labelled with a number, in a spreadsheet program.

RSS feed /ɑːesˈes ˌfiːd/ *n* A web feed format that allows subscribers to receive updates of blogs, news, podcasts, etc.

run a program /ˌrʌn ə ˈprəʊɡræm/ *v* To execute a specific program; to use a program.

S

save /seɪv/ *v* To copy information from the RAM to a storage device.

scale /skeɪl/ *v* **1** To magnify or shrink a particular font. **2** To make an object larger or smaller in any direction.

scan /skæn/ *v* To digitize an image by passing it through a scanner.

scanner /ˈskænə/ *n* An input device that scans (reads) the image as a series of dots and introduces the information into the computer's memory

screen magnifier /ˈskriːn ˌmæɡnɪfaɪə/ *n* Software that enlarges text and images on the screen, making the content more readable for users with low vision.

screen reader /ˈskriːn ˌriːdə/ *n* Software for the blind that converts screen contents into spoken words.

screensaver /ˈskriːnˌseɪvə/ *n* A program that darkens the screen after you have not worked for several minutes. Designed to protect an unchanging image from burning into the screen.

screen size /ˈskriːn ˌsaɪz/ *n* The viewing area of a monitor; measured diagonally, in inches.

scroll /skrəʊl/ *v* To move a document in its window by using scroll bars so that text in another part of the document is visible.

scroll bar /ˈskrəʊl ˌbɑː/ *n* A horizontal or vertical bar containing a box that is clicked and dragged to the desired direction.

search /sɜːtʃ/ *v* To look for specific information.

search engine /ˈsɜːtʃ ˌendʒɪn/ *n* A program that allows users to search a large database of web addresses and internet resources. Examples are Google and Yahoo!

Second Life /ˌsekənd ˈlaɪf/ *n* A 3-D virtual world on the Internet, entirely built and owned by its residents.

sector /ˈsektə/ *n* A part of a track on a magnetic disk.

seek time /ˈsiːk ˌtaɪm/ *n* The average time required for the read/write head of a disk drive to move and access data, measured in milliseconds. Also called *access time*.

set up /ˌset ˈʌp/ *v* To install and configure hardware or software.

set-top box /ˌset ˌtɒp ˈbɒks/ *n* A device that connects to a TV and to an external source of signal (e.g. a satellite dish or cable TV) and converts the signal into content then displayed on the TV screen.

setup /ˈsetʌp/ *n* The way in which a program or device is configured.

shareware /ˈʃeəweə/ *n* Software distributed similarly to freeware, but requiring payment after a trial period. Also known as 'try before you buy' software.

shopping cart /ˈʃɒpɪŋ ˌkɑːt/ *n* Software that lets you choose products from a website and processes the order through the payment gateway.

sign up /ˌsaɪn ˈʌp/ *v* To register in a service.

signature /ˈsɪɡnətʃə/ *n* A file with personal information that is automatically attached at the end of an email message.

silicon chip /ˈsɪlɪkən ˌtʃɪp/ *n* A device made up of a semi-conducting material (silicon), which contains a set of integrated circuits.

simulation /ˌsɪmjəˈleɪʃən/ *n* Using computer models (programs) to imitate real life or make predictions.

sip-and-puff /ˌsɪp ˌænd ˈpʌf/ *n* A technology that allows someone with quadriplegia to control the computer by sipping and puffing air through a mouth-controlled tube or joystick.

site /saɪt/ *n* See **website**.

Skype /skaɪp/ *n* A program that allows you to make voice and video calls from a computer.

slide scanner /ˈslaɪd ˌskænə/ *n* A device used to scan 35mm slides or film negatives; also called a *film scanner*.

smart device /ˈsmɑːt dɪˌvaɪs/ *n* An object containing a microchip and memory.

smart home /ˈsmɑːt ˌhəʊm / *n* A home where all the systems (security, lights, appliances, sensors, audio-video devices, etc.) are interconnected to allow the automatic and remote control of the home.

smart phone /ˈsmɑːt ˌfəʊn / *n* A mobile phone with advanced functions, providing voice service as well as any combination of email, text messaging, web access, voice recorder, camera, MP3, TV or video player and organizer.

smileys /ˈsmaɪliːz / *n* Faces made from punctuation characters to express emotions in email messages, e.g. :-) for happy, :-o for surprised, etc. Also called *emoticons*.

software /ˈsɒfweə/ *n* The set of program instructions that tell the computer what to do. See **hardware**.

software engineer /ˈsɑfweə endʒɪˌnɪə/ *n* Someone who writes computer programs; also known as *programmer* or *programmer analyst*.

Solaris /səˈlɑːrɪs/ *n* A Unix-based operating system, developed by Sun Microsystems, which runs on SPARC computers and other workstations.

solid modeling /ˌsɒlɪd 'mɒdəlɪŋ/ *n* A technique for representing solid objects; this includes specifying and filling the surfaces to give the appearance of a 3-D solid object with volume.

sort /sɔːt/ *v* To classify; to reorder data into a new sequence.

sound card /'saʊnd ˌkɑːd/ *n* An expansion card that processes audio signals; also called a *sound board*.

source code /'sɔːs ˌkəʊd/ *n* **1** Computer instructions written in a high-level language like C or Pascal. **2** The HTML codes of a web page.

spam /spæm/ *n* Unsolicited, junk email.

spamming /'spæmɪŋ/ *n* Posting unsolicited advertising messages.

speaker /'spiːkə/ *n* A device that provides sound output; also called a *loudspeaker*. A pair of speakers usually plug into the computer's sound card.

speech-synthesizer /ˌspiːtʃ 'sɪnθəsaɪzə/ *n* A device that produces audio output.

spell checker /'spel ˌtʃekə/ *n* A utility to correct typing mistakes.

spit /spɪt/ *n* Spam (unwanted messages) over internet telephony.

spooler /'spuːlə/ *n* A utility which makes it possible to send one document to the printer (by creating a temporary file for it) so that the user can work on another.

spreadsheet /'spredʃiːt/ *n* A program for financial planning which allows the user to analyse information presented in tabular form, by manipulating rows and columns.

spyware /'spaɪweə/ *n* A type of software that collects information from your computer without your consent.

standard toolbar /'stændəd ˌtuːlbɑː/ *n* A row of icons that, when clicked, activate certain commands of a program. For example, in a word processor, it allows you to save or print a document, include a hyperlink, check the spelling, etc.

star topology /ˌstɑː tɒ'pɒlədʒi/ *n* One of the three principal topologies for a LAN , in which all data flows through a central hub, a common connection point for the devices on the network.

storage device /'stɔːrɪdʒ dɪˌvaɪs/ *n* A hardware device used to record and store data, e.g. a hard disk, DVD or flash memory card.

store /stɔː/ *v* To copy data from the computer's internal memory to a storage device, such as a disk, tape or flash memory card.

streaming /'striːmɪŋ/ *n* A technique for transmitting sound and video so that it can be processed as a continuous stream. The files are played while they are downloading.

stylus /'staɪləs/ *n* A pen-shaped tool that is used to draw images or point to menus on pressure-sensitive screens (e.g. on PDAs).

subject /'sʌbdʒɪkt/ *n* The line that describes the content of an email.

subroutine /'sʌbruːˌtiːn/ *n* A set of instructions which performs a specific function of the program.

surf /sɜːf/ *v* To navigate and search for information on the Web.

Symbian OS /ˌsɪmbiːjən əʊ'es/ *n* An operating system used by some phone makers, including Nokia and Siemens.

system clock /'sɪstəm ˌklɒk/ *n* A clock that measures and synchronizes the flow of data.

system software /'sɪstəm ˌsɒfweə/ *n* The programs that control the basic functions of a computer, e.g. operating systems, programming software, device drivers and utilities.

T

tablet PC /ˌtæblət piː'siː/ *n* A type of notebook computer that has an LCD screen on which you can write with a stylus or digital pen. The screen can be easily folded or rotated.

telecommunications /ˌtelɪkəˌmjuːnɪ'keɪʃənz/ *n* The transmission of signals over a distance for the purpose of communication.

telegraph /'telɪgrɑːf/ *n* A communications system that transmits and receives simple electromagnetic impulses. A message transmitted by telegraph is a *telegram*.

telemarketing /'telɪˌmɑːkɪtɪŋ/ *n* The process of selling goods and services over the telephone.

teletext /'telɪtekst/ *n* A method of communicating information by using TV signals. An extra signal is broadcast with the TV picture and translated into text on the screen by a decoder.

teleworking /'telɪˌwɜːkɪŋ/ *n* The practice of working at home and communicating with the office by phone and computer. Also called *telecommuting*.

Telnet /'telnet/ *n* A protocol and a program which is used to log directly into remote computer systems. This enables you to run programs kept on them and edit files directly.

terabyte /'terəbaɪt/ *n* 1,024 gigabytes.

terminal /'tɜːmɪnəl/ *n* A hardware device, often equipped with a keyboard and a video screen, through which data can be entered or displayed.

text flow /'tekst ˌfləʊ/ *n* A feature that enables you to wrap text around images on the page.

textphone /'tekstfəʊn/ *n* A phone with a small screen and a keyboard that transcribes spoken voice as text; it is used by people with hearing or speech difficulties.

texturing /'tekstʃərɪŋ/ *n* Adding paint, colour and filters to an object in order to achieve a given look and feel.

thermal transfer printer /ˌθɜːməl 'trænsfɜː ˌprɪntə/ *n* A printer that produces colour images by adhering wax-based ink onto paper.

thesaurus /θɪ'sɔːrəs/ *n* A utility for searching synonyms and antonyms.

three-dimensional (3-D) /ˌθriːdɪ'menʃənəl/ *adj* Having three dimensions e.g. width, length, and depth. 3-D drawings represent objects more accurately.

tilt-and-swivel stand /ˌtɪlt ˌænd ˌswɪvəl 'stænd/ *n* A kind of stand that lets you move the monitor up or around, so you can use it at the right angle and height.

toner /'təʊnə/ *n* A special ink powder used in copy machines and laser printers.

toolbar /'tuːlbɑː/ *n* A row of icons on a computer screen that, when clicked, activate certain functions of a program. Toolbars are used in programs like MS Word or as add-ons for web browsers (e.g. the Google toolbar).

toolbox /'tuːlbɒks/ *n* A collection of drawing and painting tools.

topology /tɒ'pɒlədʒi/ *n* The layout or shape of a network. See **bus**, **star** and **ring topologies**.

touch screen /'tʌtʃ ˌskriːn/ *n* A display screen that is sensitive to the touch of a finger or stylus. Used in PDAs, portable game consoles, and many types of information kiosks.

touchpad /'tʌtʃˌpæd/ *n* A pointing device consisting of a soft pad which is sensitive to finger movement or pressure. Used on portable PCs.

track /træk/ *n* An area marked on the surface of a disk. When a disk is initialized, the operating system divides its surface into circular tracks, each one containing several sectors. Tracks and sectors are used to organize the information stored on disk.

trackball /'trækbɔːl/ *n* A stationary device that works like a mouse turned upside down. The ball spins freely to control the movement of the cursor on the screen. Used in laptops and CAD workstations.

translation /trænz'leɪʃən/ *n* Moving an object to a different location.

Trojan horse /ˌtrəʊdʒən 'hɔːs/ *n* Malicious software disguised as a useful program.

two-dimensional (2-D) /ˌtuːdɪ'menʃənəl/ *adj* Having only two dimensions, length and width. 2-D drawings look flat.

type style /'taɪp ˌstaɪl/ *n* A visual characteristic of a typeface, e.g. plain text, *italic*, **bold**, etc.

typeface /'taɪpˌfeɪs/ *n* The design of a set of printed characters, such as Arial and Courier. The words *typeface* and *font* are used interchangeably, but the typeface is the primary design, while the font is the particular use of a typeface, such as the size (e.g. 12 points) and style (e.g. normal, *italic*, **bold**).

U

Undo /ʌn'duː/ *n* A command that reverses or erases the last editing change done to the document.

Uniform Resource Locator (URL) /ˌjuːnɪfɔːm rɪ'zɔːs ləʊˌkeɪtə/ *n* The address of a file on the Internet, e.g. http://www.bbc.co.uk/radio.

UNIX /'juːnɪks/ *n* An operating system, designed by Bell Laboratories in the USA, found on mainframes and workstations in corporate installations.

update /ʌp'deɪt/ *v* To make something more modern or suitable for use now by adding information or changing its design.

upgradable /ʌp'greɪdəbl/ *adj* Can be upgraded or expanded.

upgrade /ʌp'greɪd/ *v* To add or replace hardware or software in order to expand the computer's power.

upload /ʌp'ləʊd/ *v* To send files to a central, often remote computer. Compare with **download**.

USB /ˌjuːes'biː/ *n* A Universal Serial Bus, a hardware interface that allows peripheral devices (disc drives, modems, cameras, etc.) to be easily connected to a computer.

USB port /juːes'biː ˌpɔːt/ *n* A USB socket on a computer device into which you can plug a USB cable.

Usenet /'juːznet/ *n* A large collection of discussion areas (called newsgroups) on the Internet.

user interface /ˌjuːzə 'ɪntəfeɪs/ *n* The standard procedures for interaction with specific computers.

user-friendly /ˌjuːzə 'frendli/ *adj* A system that is easy to learn and easy to use

username /'juːzəneɪm/ *n* **1** the part of an email address that identifies the user of the service. **2** The name you use to identify yourself when you log onto a computer system or network; also called *user ID*.

utility /juːˈtɪləti/ *n* A small program designed to improve the performance of the system. *System utility* refers to a diverse field covering anything from software designed to help you back up your hard disk or locate files, to anti-virus programs or routines used by the system.

V

Vector graphics /ˈvektə ˌgræfɪks/ *n* Images represented through the use of geometric objects such as lines, curves and polygons, based on mathematical equations. They can be changed or scaled without losing quality.

video adapter /ˈvɪdiəʊ əˌdæptə/ *n* A expansion card that processes images and sends the video signals to the monitor; also called *video graphics board*.

video editing /ˈvɪdiəʊ ˌedɪtɪŋ/ *n* The process of manipulating video images.

video projector /ˈvɪdiəʊ prəˌdʒektə/ *n* A device that projects images on a large screen using a lens system.

videoblog /ˈvɪdiəʊblɒg/ *n* A blog that includes video.

videoconferencing /ˈvɪdiəʊˌkɒnfərəntsɪŋ/ *n* A technology that allows organizations to create virtual meetings with participants in multiple locations, enabling them to talk to and see each other.

virtual interface /ˌvɜːtʃuəl ˈɪntəfeɪs/ *n* A type of interface based on virtual reality techniques. The user puts on a head-mounted display, and uses data gloves and other devices which make you feel as if you are in a 3-D world.

virtual reality /ˌvɜːtʃuəl riˈæləti/ *n* A computer-generated space in which the user interacts with artificial objects through 3-D computer simulation. This is done by using sensory peripherals, such as data gloves and head-mounted displays, to give the feeling of being immersed into an illusionary, yet sensate, world.

virus /ˈvaɪrəs/ *n* A piece of software which attaches itself to a file. Once you run an infected program, the virus quickly spreads to the system files and other software. Some viruses can destroy the contents of hard disks.

VisualBASIC /ˌvɪʒuəl ˈbeɪsɪk/ *n* A high-level programming language, developed by Microsoft in 1990, used to create graphical user interfaces in Windows applications.

VoIP /vɔɪp/ *n* Voice over Internet Protocol, which allows you to make phone calls using the Internet instead of the regular phone lines.

voice recognition /ˌvɔɪs rekəgˈnɪʃən/ *n* A technology that allows computers to interpret human speech, converting spoken words into digitized text or instructions.

VoiceXML /ˌvɔɪs eksemˈel/ *n* A markup language which makes web content accessible via voice and phone. Short for *Voice Extensible Markup Language*.

volatile memory /ˌvɒlətaɪl ˈmeməri/ *n* Temporary memory (e.g. RAM); it doesn't hold its contents without power.

W

wearable computer /ˌweərəbəl kəmˈpjuːtə/ *n* A computer that is worn on the body, or integrated into the user's clothing.

Web /web/ *n* A network of documents that works in a hypertext environment, i.e. using text that contains links to other documents. It's also known as the *World Wide Web*, WWW or W3. By using a special program known as a browser, you can find information on nearly any topic you can imagine.

Web Accessibility Initiative (**WAI**) /ˌweb əksesəˈbɪləti ɪˌnɪʃətɪv/ *n* A project that tries to make the Web accessible to people with disabilities.

web editor /ˈweb ˌedɪtə/ *n* Software that lets you design web pages without writing HTML codes.

web page /ˈweb ˌpeɪdʒ/ *n* An individual document on the Web, identified by its own unique URL. Web pages contain different elements, such as text, pictures, video, links, etc.

webcam /ˈwebkæm/ *n* A web camera used to send live video images via the Internet.

webcasting /ˈwebkɑːstɪŋ/ *n* Sending audio and video live over the Internet.

webmaster /ˈwebˌmɑːstə/ *n* Someone responsible for designing, developing, marketing or maintaining websites.

website /ˈwebsaɪt/ *n* A collection of web pages (usually including a homepage), set up by an organization or an individual, which are usually stored on the same server. The pages are all linked together; you can move from one page to another by clicking on words or pictures called *hyperlinks*.

wide area network (WAN) /ˌwaɪd ˌeəriə ˈnetwɜːk/ *n* A network that extends outside a building or small area. For long distance communications, LANs are usually connected into a WAN. The largest WAN is the Internet.

Wi-Fi /ˈwaɪfaɪ/ *n* A term from the Wi-Fi Alliance, which certifies that network devices comply with the IEEE 802.11 wireless specifications. A typical Wi-Fi setup contains one or more wireless access points (base stations) and various computer devices acting as clients.

Wi-Fi phone /ˌwaɪfaɪ ˈfəʊn/ *n* A mobile phone that can switch from the cellular network to a wireless VoIP network and vice versa.

Wii /wiː/ *n* A video game console from Nintendo, introduced in 2006.

wiki /ˈwɪkiː/ *n* A collaborative website whose content can be edited by anyone who has access to it, e.g. WikiWikiWeb, Wikipedia, etc.

WiMAX /ˈwaɪmæks/ *n* A technology that enables the delivery of wireless broadband access as an alternative to cable and ADSL; short for *Worldwide Interoperability for Microwave Access*.

window /ˈwɪndəʊ/ *n* A scrollable viewing area on screen, which can contain files or folders.

Windows /ˈwɪndəʊz/ *n* The operating system from Microsoft that runs on most PCs. The most recent versions are Windows 2000, Windows XP and Windows Vista.

Windows Mobile /ˌwɪndəʊz ˈməʊbaɪl/ *n* An operating system used on many PDAs and smartphones.

Windows Vista /ˌwɪndəʊz ˈvɪstə/ *n* The new Windows, released in 2007. It includes security improvements, a new graphical user interface, and new ways of searching information

wired /waɪəd/ *adj* Equipped with a system of wires (cables).

wireframe /ˌwaɪəˈfreɪm/ *n* The drawing of a model by tracing features like edges or contour lines.

wireless /ˈwaɪələs/ *adj* Having no wires; without the use of cables.

Wireless access point (WAP) /ˌwaɪələs ˈækses ˌpɔɪnt/ *n* A device that connects wireless communication devices together to form a wireless network.

wireless adapter /ˌwaɪələs əˈdæptə/ *n* A device that adds wireless connectivity to a computer or PDA. It is attached via a PC card or a USB port. There are three main types of wireless adapters: Bluetooth, cellular (for mobiles) and Wi-Fi (for laptops and desktop PCs).

wireless LAN /ˌwaɪələs ˈlæn/ *n* A wireless local area network, linking two or more computers without cables.

wireless network /ˌwaɪələs ˈnetwɜːk/ *n* Any type of network that uses electromagnetic waves, such as radio waves, to transmit data. These are the main types: satellites for long distances, WiMAX for connecting Wi-Fi hotspots, Wi-Fi for medium-range distances, Bluetooth for short distances, and GSM for mobile phones.

wireless router /ˌwaɪələs ˈruːtə/ *n* A device which allows computers to communicate via radio waves. Also called *wireless access point* or *base station*.

Word /wɜːd/ *n* A word processor from Microsoft.

word processor /ˈwɜːd ˌprəʊsesə/ *n* An application that manipulates text and produces documents suitable for printing.

word wrap /ˈwɜːd ˌræp/ *n* An editing facility which automatically moves a word to the next line if there is not enough space for the complete word on the current line.

workstation /ˈwɜːkˌsteɪʃən/ *n* **1** A high-performance computer, typically used for graphics, CAD, software development and scientific applications. **2** Any computer connected to a network.

World Wide Web /ˌwɜːld ˌwaɪd ˈweb/ *n* See **Web**.

worm /wɜːm/ *n* A self-copying program that spreads through email attachments; it replicates itself and sends a copy to everyone in a contact list.

X

Xbox 360 /ˌeksbɒks ˌθriː ˈsɪksti/ *n* A video game console from Microsoft.

XML /ˌeksemˈel/ *n* Extensible Markup Language. While HTML uses pre-defined tags, XML allows us to create our own tags to better describe data.

Y

Yahoo! /jɑːˈhuː/ *n* A leading web portal, with a mix of news, entertainment and online shopping, as well as search engine, internet directory, email and IM services.

YouTube /ˈjuːtjuːb/ *n* A popular website which lets users upload, view, and share video clips.

Z

.zip /zɪp/ *n* An extension that identifies compressed files. To decompress them you need a shareware program like WinZip.

zoom /zuːm/ *n* A tool used to magnify areas of an image when you are doing close, detailed work.

Irregular verbs

These are the most important irregular verbs. They can be divided into the following groups (A–E):

A All three forms the same

Base	Past simple	Past participle	*Translation*
bet	bet	bet	....
cost	cost	cost	....
cut	cut	cut	....
hit	hit	hit	....
hurt	hurt	hurt	....
set	set	set	....
let	let	let	....
put	put	put	....
shut	shut	shut	....
spread	spread	spread	....
read	read /red/	read /red/	....

B Base = Past simple

Base	Past simple	Past participle	*Translation*
beat	beat	beaten	....

C Past simple = Past participle

Base	Past simple	Past participle	*Translation*
bend	bent	bent	....
bleed	bled	bled	....
bring	brought	brought	....
build	built	built	....
buy	bought	bought	....
catch	caught	caught	....
deal	dealt	dealt	....
feed	fed	fed	....
feel	felt	felt	....
fight	fought	fought	....
find	found	found	....
get	got	got	....
hang	hung	hung	....
have	had	had	....

C Past simple = Past participle cont.

Base	Past simple	Past participle	*Translation*
hear	heard	heard	....
hold	held	held	....
keep	kept	kept	....
lay	laid	laid	....
lead	led	led	....
learn	learnt	learnt	....
leave	left	left	....
lend	lent	lent	....
light	lit	lit	....
lose	lost	lost	....
make	made	made	....
mean	meant	meant	....
meet	met	met	....
pay	paid	paid	....
say	said	said	....
sell	sold	sold	....
send	sent	sent	....
shine	shone	shone	....
shoot	shot	shot	....
sit	sat	sat	....
sleep	slept	slept	....
spend	spent	spent	....
stand	stood	stood	....
stick	stuck	stuck	....
strike	struck	struck	....
sweep	swept	swept	....
teach	taught	taught	....
tell	told	told	....
think	thought	thought	....
understand	understood	understood	....
win	won	won	....

D Base = Past participle

Base	Past simple	Past participle	*Translation*
become	became	become	…………
come	came	come	…………
run	ran	run	…………

E All three forms different

Base	Past simple	Past participle	*Translation*
arise	arose	arisen	…………
awake	awoke	awoken	…………
be	was/were	been	…………
begin	began	begun	…………
bite	bit	bitten	…………
blow	blew	blown	…………
break	broke	broken	…………
choose	chose	chosen	…………
do	did	done	…………
draw	drew	drawn	…………
drink	drank	drunk	…………
drive	drove	driven	…………
eat	ate	eaten	…………
fall	fell	fallen	…………
fly	flew	flown	…………
forbid	forbade	forbidden	…………
forget	forgot	forgotten	…………
forgive	forgave	forgiven	…………
freeze	froze	frozen	…………

E All three forms different cont.

Base	Past simple	Past participle	*Translation*
give	gave	given	…………
go	went	gone	…………
grow	grew	grown	…………
hide	hid	hidden	…………
know	knew	known	…………
lie	lay	lain	…………
overwrite	overwrote	overwritten	…………
ride	rode	ridden	…………
ring	rang	rung	…………
rise	rose	risen	…………
see	saw	seen	…………
shake	shook	shaken	…………
show	showed	shown	…………
shrink	shrank	shrunk	…………
sing	sang	sung	…………
sink	sank	sunk	…………
speak	spoke	spoken	…………
steal	stole	stolen	…………
swear	swore	sworn	…………
swim	swam	swum	…………
take	took	taken	…………
tear	tore	torn	…………
throw	threw	thrown	…………
wake	woke	woken	…………
wear	wore	worn	…………
withdraw	withdrew	withdrawn	…………
write	wrote	written	…………

Acronyms and abbreviations

ADSL Asymmetric Digital Subscriber Line
AI Artificial Intelligence
AIM AOL Instant Messenger
ALU Arithmetic Logic Unit
AMD Advanced Micro Devices
ASCII American Standard Code for Information Interchange
AT&T American Telephone & Telegraph company
ATA Analogue Telephone Adaptor
ATM Automated Teller Machine
AVI Audio Video Interface

BASIC Beginner's All-purpose Symbolic Instruction Code
BBS Bulletin Board System
Bcc: Blind carbon (or courtesy) copy
BIOS Basic Input/Output System
bit binary digit
bps bits per second

CAD Computer-Aided Design
Cc: Carbon (or courtesy) copy
CCD Charge-Coupled Devices
CD Compact Disc
cd/m2 Candela per square metre
CD-R Compact Disc-Recordable
CD-ROM Compact Disc-Read Only Memory
CD-RW Compact Disc-Rewritable
CERN Conseil Européen pour la Recherche Nucléaire
COBOL COmmon Business-Oriented Language
CPU Central Processing Unit
CRT Cathode Ray Tube
CSS Cascading Style Sheets
CTP Computer To Plate
CU Control Unit

DAB Digital Audio Broadcasting
DAW Digital Audio Workstation
DBMS Database Management System
DDR Double Data Rate (RAM)
DIMM Dual In-line Memory Module
DLP Digital-Light processing
DMB Digital Multimedia Broadcasting
DNS Domain Name System
dpi dots per inch
DTP Desktop Publishing
DTTV Digital Terrestrial television
DVB-H Digital Video Broadcast-Handheld
DVD-/+RW Digital Versatile Disc-Rewritable
DVD Digital Versatile Disc or Digital Video Disc
DVD-R Digital Versatile Disc-Recordable
DVD-ROM Digital Versatile Disc-Read Only Memory
DVI Digital Video Interface

EEPROM Electrically Erasable Programmable ROM
EPS Encapsulated PostScript

FAQ Frequently Asked Questions
FORTRAN FORmula TRANslation
FTP File Transfer Protocol

GB Gigabyte (1,024 megabytes)
GHz Gigahertz
GIF Graphic Interchange Format
GIS Geographic Information System
GNU Gnu's Not UNIX
GPS Global Positioning System
GSM Global System for Mobile communication
GUI Graphical User Interface

HDD Hard Disk Drive
HD-DVD High Definition-Digital Versatile Disk
HDTV High-definition 'Television
HP Hewlett-Packard
HTML Hypertext Markup Language
HTTP Hypertext Transfer Protocol
Hz Hertz

I/O Input/Output
IBM International Business Machines
ICQ I Seek You
ICT Information and Communications Technologies
IM Instant Messaging
IP Internet Protocol
IR Instruction Register
IrDA Infrared Data Association
ISP Internet Service Provider
IT Information technology

JPG (or JPEG) Joint Photographic Experts Group

k 1 kilo, used to denote a thousand; 2 1,024 bytes
KB kilobyte (1,024 bytes)

LAN Local Area Network
Laser Light Amplification by Stimulated Emission of Radiation
LCD Liquid-Crystal Display
LISP LISt Processing

.mov QuickTime movie
Mac Macintosh computer
MAN Metropolitan Area Network
MB Megabyte (1,024 kilobytes)
MHz Megahertz
MIDI Musical Instrument Digital Interface
MIPS Million Instructions Per Second
MMS Multimedia messages
Modem MOdulator/DEModulator
MP3 MPEG-1 Layer-3 Audio
MPEG Moving Pictures Experts Group
ms millisecond

NIC Network Interface Card
NUI Network User Identifier

OCR Optical Character Recognition
OLE Microsoft's Object Linking and Embedding standard
OLED Organic Light-Emitting Diodes (display)
OOP Object Oriented Programming
OS Operating System

. pdf portable document format
PAN Personal Area Network
PC 1Personal Computer; 2 Program Counter
PCL Printer Control Language
PDA Personal Digital Assistant
PDL Page Description Language
PGP Pretty Good Privacy
PIN Personal Identification Number
pixel picture element
png portable network graphic
ppm pages per minute
PPP Point to Point Protocol

.ra RealAudio file
RAM Random Access Memory
RGB Red, Green, Blue
RFID Radio-Frequency identification
RIM Research In Motion
RIP Raster Image Processor
RISC Reduced Instruction Set Computer
ROM Read Only Memory
rpm revolutions per minute
RSI repetitive strain injury
RSS Really Simple Syndication or Rich Site Summary

SDRAM Synchronous Dynamic Random Access Memory
SIM (card) Subscriber Identity Module
SMS Short Message Service
SMTP Simple Mail Transfer Protocol
SQL Structured Query Language
SSL Secure Sockets Layer
SXGA Super XGA (Extended Graphics Array)

TAN Transaction Authorization Number
TB Terabyte (1,024 gigabytes)
TCP/IP Transmission Control Protocol / Internet Protocol
TFT Thin Film Transistor (display)
TIFF Tagged Image File Format

UMTS Universal Mobile Telecommunications System
URL Uniform Resource Locator
USB Universal Serial Bus

VAT Value Added Tax
VCR Videocassette Recorder
VDU Visual Display Unit
VGA Video Graphics Adapter/Array
VoiceXML Voice Extensible Markup Language
VoIP Voice over Internet Protocol
VRML Virtual Reality Modelling (or Markup) Language

.wav Windows wave audio file
W3 See **Web** in Glossary
WAI Web Accessibility Initiative
WAN Wide Area Network
WAP 1 wireless access point; 2 Wireless Application Protocol
Wi-Fi Wireless Fidelity
WiMAX Worldwide Interoperability for Microwave Access
WIMP Window, Icon, Menu (or mouse) and Pointer
WP Word Processing
WWW World Wide Web
WYSIWYG What You See Is What You Get

XGA Extended Graphics Array
XML Extensible Markup Language
WXGA Wide XGA (Extended Graphics Array)

Infotech

English for computer users

Student's Book Fourth Edition

Santiago Remacha Esteras

Now in its fourth edition, **Infotech** is a comprehensive course in the English of computing, used and trusted by students and teachers all over the world.

Fully updated in line with the latest developments in Information Communications Technology, this edition teaches intermediate students the language and skills they need to understand and work in the world of computers. The 30 topic-based units cover everything from computer essentials through to programming, web design, job hunting and future technologies.

A focus on terminology is combined with vocabulary and grammar practice to give students the tools to use English in areas such as describing features and functions, chatting online, applying for jobs and discussing the world of ICT.

With the support of clear explanations, no specialist knowledge of ICT is required, making this course ideal for anyone who needs to understand the English of computing for study or work.

Key features

- Technical reading texts and realistic listening material keep learners up-to-date with recent developments in the fast-moving world of computing, from Windows Vista to Blu-ray
- A comprehensive glossary of computer terms and acronyms provides valuable support
- A modular structure enables teachers to focus on the most relevant topics and language for their students
- New interactive internet research tasks and an online workbook enable students to use their ICT skills as they learn

www.cambridge.org/elt/ict
www.klett.de

Learning Links

We recommend for use with **Infotech** Fourth Edition:

ISBN: 978 3 12 539585 5

ISBN: 978 3 12 534086 2

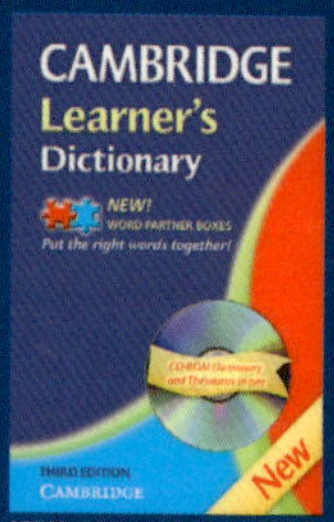

ISBN: 978 3 12 517959 2

CAMBRIDGE
UNIVERSITY PRESS
www.cambridge.org

ISBN 978-3-12-534261-3
9 783125 342613